# OPPORTUNITIES TO IMPROVE THE U.S. GEOLOGICAL SURVEY NATIONAL WATER QUALITY ASSESSMENT PROGRAM

Committee to Improve the U.S. Geological Survey
National Water Quality Assessment Program

Water Science and Technology Board

Division on Earth and Life Studies

National Research Council

NATIONAL ACADEMY PRESS
Washington, D.C.

NATIONAL ACADEMY PRESS • 2101 Constitution Avenue, N.W. • Washington, DC 20418

NOTICE: The project that is the subject of this report was approved by the Governing Board of the National Research Council, whose members are drawn from the councils of the National Academy of Sciences, the National Academy of Engineering, and the Institute of Medicine. The members of the committee responsible for the report were chosen for their special competences and with regard for appropriate balance.

Support for this project was provided by the U.S. Geological Survey under Grant No. 99HQAG0184.

International Standard Book Number: 0-309-08305-2
Library of Congress Card Number: 2002101389

Additional copies of this report are available from:
National Academy Press
2101 Constitution Avenue, N.W.
Box 285
Washington, DC 20055
800-624-6242
202-334-3313 (in the Washington metropolitan area)
http://www.nap.edu

Copyright 2002 by the National Academy of Sciences. All rights reserved.

Printed in the United States of America.

# THE NATIONAL ACADEMIES

National Academy of Sciences
National Academy of Engineering
Institute of Medicine
National Research Council

The **National Academy of Sciences** is a private, nonprofit, self-perpetuating society of distinguished scholars engaged in scientific and engineering research, dedicated to the furtherance of science and technology and to their use for the general welfare. Upon the authority of the charter granted to it by the Congress in 1863, the Academy has a mandate that requires it to advise the federal government on scientific and technical matters. Dr. Bruce M. Alberts is president of the National Academy of Sciences.

The **National Academy of Engineering** was established in 1964, under the charter of the National Academy of Sciences, as a parallel organization of outstanding engineers. It is autonomous in its administration and in the selection of its members, sharing with the National Academy of Sciences the responsibility for advising the federal government. The National Academy of Engineering also sponsors engineering programs aimed at meeting national needs, encourages education and research, and recognizes the superior achievement of engineers. Dr. Wm. A. Wulf is president of the National Academy of Engineering.

The **Institute of Medicine** was established in 1970 by the National Academy of Sciences to secure the services of eminent members of appropriate professions in the examination of policy matters pertaining to the health of the public. The Institute acts under the responsibility given to the National Academy of Sciences by its congressional charter to be an adviser to the federal government and, upon its own initiative, to identify issues of medical care, research, and education. Dr. Kenneth I. Shine is president of the Institute of Medicine.

The **National Research Council** was organized by the National Academy of Sciences in 1916 to associate the broad community of science and technology with the Academy's purposes of furthering knowledge and advising the federal government. Functioning in accordance with general policies determined by the Academy, the Council has become the principal operating agency of both the National Academy of Sciences and the National Academy of Engineering in providing services to the government, the public, and the scientific and engineering communities. The Council is administered jointly by both Academies and the Institute of Medicine. Dr. Bruce M. Alberts and Dr. Wm. A. Wulf are chairman and vice chairman, respectively, of the National Research Council.

**COMMITTEE TO IMPROVE THE U.S. GEOLOGICAL SURVEY NATIONAL WATER QUALITY ASSESSMENT PROGRAM**

GEORGE R. HALLBERG, *Chair,* The Cadmus Group, Inc., Waltham, Massachusetts
MICHAEL E. CAMPANA, University of New Mexico, Albuquerque
DANIEL B. CARR, George Mason University, Fairfax, Virginia
LORRAINE L. JANUS, New York City Department of Environmental Protection, Valhalla, New York
JUDITH L. MEYER, University of Georgia, Athens
KENNETH H. RECKHOW, Duke University, Durham, North Carolina
MARC O. RIBAUDO, Economic Research Service, U.S. Department of Agriculture, Washington, D.C.
PAUL V. ROBERTS, Stanford University, California *(until August 2000)*
KENNETH K. TANJI, University of California, Davis
RICHARD M. VOGEL, Tufts University, Medford, Massachusetts
MARYLYNN V. YATES, University of California, Riverside

*Staff*

MARK C. GIBSON, Study Director
LAURA H. EHLERS, Study Director *(until December 2000)*
ELLEN A. DE GUZMAN, Research Associate

## WATER SCIENCE AND TECHNOLOGY BOARD

RICHARD G. LUTHY, *Chair*, Stanford University, Stanford, California
JOAN B. ROSE, *Vice-Chair*, University of South Florida, St. Petersburg
RICHELLE M. ALLEN-KING, Washington State University, Pullman
GREGORY B. BAECHER, University of Maryland, College Park
KENNETH R. BRADBURY, Wisconsin Geological and Natural History Survey, Madison, Wisconsin
JAMES CROOK, CH2M Hill, Boston, Massachusetts
EFI FOUFOULA-GEORGIOU, University of Minnesota, Minneapolis
PETER GLEICK, Pacific Institute, Oakland, California
STEVEN P. GLOSS, U.S. Geological Survey, Flagstaff, Arizona
JOHN LETEY, JR., University of California, Riverside
DIANE M. MCKNIGHT, University of Colorado, Boulder
CHRISTINE L. MOE, Emory University, Atlanta, Georgia
RUTHERFORD H. PLATT, University of Massachusetts, Amherst
JERALD L. SCHNOOR, University of Iowa, Iowa City
LEONARD SHABMAN, Virginia Polytechnic Institute and State University, Blacksburg
R. RHODES TRUSSELL, Montgomery Watson Harza, Pasadena, California

*Staff*

STEPHEN D. PARKER, Director
LAURA J. EHLERS, Senior Staff Officer
JEFFREY W. JACOBS, Senior Staff Officer
WILLIAM S. LOGAN, Senior Staff Officer
MARK C. GIBSON, Staff Officer
M. JEANNE AQUILINO, Administrative Associate
PATRICIA JONES KERSHAW, Study/Research Associate
ELLEN A. DE GUZMAN, Research Associate
ANITA A. HALL, Administrative Assistant
ANIKE L. JOHNSON, Project Assistant
JON Q. SANDERS, Project Assistant

# Preface

From the 1950s through the 1970s, events ranging from health advisories to burning rivers brought public attention and scrutiny to the deteriorating state of water quality in the United States. Local, state, and even regional monitoring data confirmed problems in many areas. Such information contributed to the evolution of the environmental movement and the development of the Clean Water Act and the Safe Drinking Water Act. It also became clear that the nation had no program that was providing an adequate or systematic monitoring network to assess the status of waters throughout the nation, particularly related to the contaminants that had become the major concern in this modern era. Some federal, state, and even local organizations had developed excellent water quality monitoring programs, but they were often focused solely on local problems, used different methods, or were looking at geochemical phenomena, not the contaminant issues of real concern. As such, they could not adequately be amalgamated to provide a national picture. In the 1980s, Congress, federal and state agencies, and industry began to call for development of national approaches to assess and track water quality and to answer such fundamental questions as, Is the quality of water across the nation getting better or worse?

Since its inception in the late 1800s, the U.S. Geological Survey (USGS) has become a major national contributor of scientific investigations and information about the nation's waters. In particular, the USGS has long provided national and scientific leadership in the understanding of surface water hydrology, runoff, flood and discharge studies, water quantity issues, and more recently, groundwater hydrology. In the 1980s, the USGS was challenged to expand this role to address the quality of the nation's waters, and it responded with the development of the National Water Quality Assessment (NAWQA) Program. In fiscal year 1986,

Congress appropriated funds for the establishment of the NAWQA pilot program. NAWQA has since evolved from a pilot program into a respected, mature national monitoring program of unprecedented scope—one for which hopes and expectations also run high. From its earliest concept to the current plans for the future, three goals drive NAWQA's design and development: (1) *status*—to provide a nationally consistent description of the current water quality conditions for a large part of the nation's water resources; (2) *trends*—to define long-term trends (or lack of trends) in water quality; and (3) *understanding*—to identify, describe, and explain (to the extent possible) the major factors that affect (and cause) observed water quality conditions and trends. Although the exact wording of the goals has been refined over time, these three goals are the organizing themes for NAWQA's past, present, and future.

Designing and implementing a national water quality assessment program is obviously a challenging task. Throughout the history of NAWQA, the USGS has periodically requested input and assistance from the Water Science and Technology Board (WSTB) of the National Research Council (NRC) to provide programmatic reviews of NAWQA or its key components—a reflection of the USGS's ongoing concern to ensure the quality of scientific approaches for this important program. In 1999, the NRC was again approached to have WSTB convene a committee of experts to assist the continued development of NAWQA.

For this study, the committee was charged to provide guidance to the USGS on opportunities to improve the NAWQA program. More specifically, the committee was to conduct an initial assessment of general accomplishments in the NAWQA program to date by engaging in discussions with program scientists and others such as users of NAWQA products and by reviewing USGS internal reports on opportunities to improve NAWQA. The statement of task notes that the four main activities of the study committee was then to: (1) recommend methods for the improved understanding of the causative factors affecting water quality conditions; (2) determine whether information produced in the program can be extrapolated so as to allow inferences about water quality conditions in areas not studied intensely in NAWQA; (3) assess the completeness and appropriateness of priority issues (e.g., pesticides, nutrients, volatile organic compounds, trace elements) selected for broad investigation under the national synthesis component of the program; and (4) describe how information generated at the study unit scale can be aggregated and presented so as to be meaningful at the regional and national levels.

Though not specifically noted in the committee's statement of task, the timing of this latest request carried with it an imperative implicit in its initial charge to "provide guidance to the U.S. Geological Survey on opportunities to improve the NAWQA program." As the committee was being formed, NAWQA was completing its first decade of extensive nationwide monitoring (called Cycle I) and refining plans for its second decade (Cycle II). Cycle I focused on status assessments (also called occurrence and distribution assessments) that define the condi-

PREFACE ix

tion of the water resources throughout the nation. Defining status is an important and sometimes daunting task by itself, but Cycle II must now move beyond and build upon these status assessments to begin assessment of water quality trends and to further our scientific understanding of the why and how behind water quality status and trends. These goals are integral to the grand design of NAWQA and largely reflect the charge that policy makers have placed on NAWQA since its inception. As noted in this report, if there are substantive opportunities to improve NAWQA, then this is an ideal time to do so as this new round of monitoring and studies is about to begin.

The members of this committee brought a wide range of water resources expertise and a range of experience in interacting with NAWQA that made for enlivened and enlightening discussion throughout and ultimately led us to the forward-looking recommendations contained herein. Some members have had associations with NAWQA since its very inception, including service on earlier NRC committees' reviewing NAWQA (which provided institutional memory and perspective); other members were primarily users and consumers of NAWQA data and reports. The committee held five deliberative meetings; at four of these meetings the committee heard presentations from and engaged "in discussions with program scientists and others such as users of NAWQA products," as required in its statement of task. The committee did so, not just to "conduct an initial assessment of general accomplishments," but also to gather testimony and insight on where the opportunities existed to improve NAWQA. Throughout the course of the study, committee members personally visited with NAWQA staff, particularly field (study unit) staff; other USGS (non-NAWQA) personnel, other local, state, and federal agency "users" of NAWQA data and information; and other research users, casting a wide net for input to the deliberative process. The committee also collectively reviewed scores of NAWQA-related reports—both as users and to support these deliberations.

The committee thanks several persons external to the USGS who nonetheless provided highly informative and useful presentations regarding their collective experiences with the NAWQA program at the first three committee meetings, including: Margarete Heber, U.S. Environmental Protection Agency; Emery Cleaves, Maryland Geological Survey; Greg Woodside, Orange County (California) Water District; Karen Schaffer, New Jersey Department of Environmental Protection; and Jim Reilly, New Jersey Office of State Planning. We would particularly like to thank the USGS NAWQA staff as a whole, particularly Tim Miller, Bill Wilber, Bob Gilliom, and Carol Couch, for answering our many inquiries and requests for reports and documents. The committee also thanks the NRC WSTB staff for their support and leadership. When the study was initiated, Laura Ehlers was the study director. When she had to move to other responsibilities in December 2000, Mark Gibson assumed the study director post. We thank them both for their support and leadership. In particular, we thank Mark for his significant contributions to the report and efforts to bring the study to completion. Ellen de Guzman, research associate with the WSTB, provided excellent staff support throughout the study. We also thank former committee member Paul Roberts of Stanford University for his past insights and contributions, many of which carried over into this report.

This report has been reviewed in draft form by individuals chosen for their diverse perspectives and technical expertise in accordance with procedures approved by the NRC's Report Review Committee. The purpose of this independent review is to provide candid and critical comments that will assist the institution in making its published report as sound as possible and to ensure that the report meets institutional standards for objectivity, evidence, and responsiveness to the study charge. The review comments and draft manuscript remain confidential to protect the integrity of the deliberative process. We wish to thank the following individuals for their review of this report: Susan Davies, Maine Department of Environmental Protection; Don Epp, Pennsylvania State University; Daniel Loucks, Cornell University; David Moreau, University of North Carolina; Don Siegel, Syracuse University; Kent Thornton, FTN & Associates; and Robert Ward, Colorado Water Resources Research Institute.

Although the reviewers listed above have provided many constructive comments and suggestions, they were not asked to endorse the conclusions or recommendations nor did they see the final draft of the report before its release. The review of this report was overseen by Henry J. Vaux, University of California. Appointed by the National Research Council, he was responsible for making certain that an independent examination of this report was carried out in accordance with institutional procedures and that all review comments were carefully considered. Responsibility for the final content of this report rests entirely with the authoring committee and the institution.

The committee hopes that this report will help to strengthen NAWQA both as it enters its second decade of nationwide water quality monitoring and assessment and, hopefully, beyond. NAWQA has already become the premier program assessing status and trends of water quality at a national level. Yet the program strives to continue to improve its efficiency, visibility, and above all, utility, which the committee strongly supports and encourages. Scientists, policy makers, and legislative leaders must recognize that identifying and truly understanding water quality status and trends is a long-term undertaking, requiring long-term support. Even when the answers are unexpected or not popular, such programs must be able to count on long-term and adequate support to evaluate the ultimate performance of policy decisions and the effects of society on our important water resources.

GEORGE R. HALLBERG
*Chair*, Committee to Improve the U.S. Geological Survey
National Water Quality Assessment Program

# Contents

EXECUTIVE SUMMARY 1

1 INTRODUCTION 17
Overview of Cycle I of NAWQA, 22
Initial Assessment of NAWQA Results and Accomplishments, 31
Introduction to Cycle II of NAWQA, 37
Conclusions, 39
References, 40

2 TRANSITION FROM CYCLE I TO CYCLE II:
REPRESENTATIVENESS OF STUDY UNITS 42
Introduction, 42
Representativeness of the Study Unit Approach, 46
Lakes, Reservoirs, Coastal Waters, and Permafrost, 53
Surface Water and Groundwater, 56
Conclusions and Recommendations, 57
References, 58

3 NAWQA CYCLE II GOALS—STATUS 61
Introduction, 61
Resources Not Previously Sampled and Drinking Water Sources, 62
Contaminants Not Previously Sampled, 68
Importance of Conducting Sediment Monitoring, 79
Conclusions and Recommendations 86
References, 89

| | | |
|---|---|---|
| 4 | NAWQA CYCLE II GOALS—TRENDS AND STATISTICAL SUPPORT FOR UNDERSTANDING | 93 |

Introduction, 93
Design of Water Quality Monitoring Networks and Programs, 95
Evaluation of Trends in Water Quality, 100
Nonexperimental Approaches to Enhance Scientific Understanding, 104
Effects of Urbanization, 105
Response to Agricultural Management Practices, 110
Conclusions and Recommendations, 115
References, 117

| | | |
|---|---|---|
| 5 | NAWQA CYCLE II GOALS—UNDERSTANDING | 121 |

Introduction, 121
Role of Models in Understanding Cause and Effect, 124
Proposed Implementation Approach for the Understanding Goal of Cycle II, 131
Themes and Objectives of Cycle II Understanding Goal, 133
Conclusions and Recommendations, 153
References, 158

| | | |
|---|---|---|
| 6 | COMMUNICATING NAWQA DATA AND INFORMATION TO USERS | 162 |

Introduction, 162
Information Communicated by NAWQA, 163
Methods of Communicating Results, 165
Policy Relevance of NAWQA, 171
Conclusions and Recommendations, 176
References, 178

| | | |
|---|---|---|
| 7 | COOPERATION AND COORDINATION ISSUES | 182 |

Introduction, 182
NAWQA Liaison Committees, 184
USGS District Programs, 185
Other USGS Water Resources Research Programs, 186
Cooperation with Other Agencies, 188
NAWQA and the States, 191
Conclusions and Recommendations, 198
References, 201

| | | |
|---|---|---|
| 8 | THE FUTURE OF NAWQA | 203 |

Four Crosscutting Issues, 205
NAWQA—Past, Present, and Future, 210
NAWQA and *Future Roles and Opportunities for the USGS,* 211

## APPENDIXES
A   Extracts from *Study-Unit Design Guidelines for Cycle II of the National Water Quality Assessment (NAWQA)*   217
B   Biographical Information   234

# Executive Summary

The U.S. Geological Survey (USGS) has historically been regarded as the primary national source of information on water quantity and quality in the United States, including both surface waters and groundwater. Building on this history, the USGS developed the National Water Quality Assessment (NAWQA) Program in 1985 to assess past, current, and future water quality conditions and trends in representative river basins and aquifers across the United States. Fundamental to NAWQA's design was an integrated monitoring network using highly consistent methods. More specifically, the three goals of the program are to address the following: (1) *status*—description of water quality conditions for a large representative part of the nation's freshwater resources; (2) *trends*—assessment of long-term changes in the quality of water resources; and (3) *understanding*—analysis of how human-related activities, management strategies, and the natural environment interact to control water quality in different parts of the nation. Although the exact wording of these goals has been refined over time, they continue to be the organizing themes for NAWQA's future.

During its first decade of extensive monitoring (1991 to 2001) known as Cycle I, NAWQA concentrated primarily on gathering comparable information on water quality ("status assessments") in dozens of geographic areas called study units that include major river basins and/or aquifers nationwide. Perhaps the most important facet of the program is the similar design of each investigation and the use of standardized methods that make comparisons among disparate study units possible. Comparing and analyzing data from individual study units has led to regional and national assessments of water quality, collectively referred to as "national synthesis."

Nearing completion of Cycle I, USGS scientists are planning for the program's future and have requested the input of the National Research Council (NRC) to help shape NAWQA activities during the program's second decade of monitoring, called Cycle II. In this regard, a major concern of the USGS is to maximize program usefulness for a wide variety of decision makers, managers, and planners at all levels of government, as well as nongovernmental organizations, industry, academia, and the public sector. Indeed, data and information from the NAWQA program can be integral to research, monitoring, and regulatory activities at the local, state, and regional levels. In Cycle II, NAWQA will devote more resources to studying trends in water quality and understanding the factors that cause changes in water quality and fewer resources to water quality status assessments that comprised the majority of work in Cycle I. The USGS seeks NRC guidance in planning and executing this ambitious agenda as described by the committee's statement of task, which is summarized below.

The overarching request to the committee was "to provide guidance to the U.S. Geological Survey on opportunities to improve the NAWQA program." The USGS specifically asked the committee to begin with an initial assessment of the general accomplishments of the NAWQA program to date. Next, it asked the committee to focus on four particular areas of NAWQA as it enters Cycle II: (1) suggest methods to improve understanding of the causative factors affecting water quality conditions; (2) assess whether information produced in the program can be extrapolated to allow inferences about water quality conditions in areas not studied intensely under NAWQA; (3) examine current priority issues (i.e., pesticides, nutrients, volatile organic compounds, trace elements, and ecological synthesis) selected for broad investigation under NAWQA for completeness; and (4) make recommendations on aggregation and presentation of information generated at the study unit scale so that it is meaningful at the regional and national levels.

The USGS has sought advice from the NRC throughout the evolution of the NAWQA program. The NRC's Water Science and Technology Board (WSTB) has provided advice to USGS regarding NAWQA four separate times in the past as the program has evolved from an unfunded concept in 1985 to a relatively mature and established program in 2001. The origin and relationship of these four reports and their major conclusions and recommendations are summarized briefly in Chapter 1 of this report. This report was written by the Committee to Improve the U.S. Geological Survey National Water Quality Assessment Program, comprised of 10 (originally 11) experts in hydrology, hydrogeology and geochemistry, water quality monitoring and modeling, aquatic ecology, public health microbiology, environmental economics, and statistics and information technology. The contents, conclusions, and recommendations of this report are based on a review of relevant technical literature, information gathered at and between five committee meetings, and the expertise of committee members. Furthermore, because of space limitations, this Executive Summary includes only the major conclusions and related recommendations of the committee in the

## EXECUTIVE SUMMARY

general order of their appearance in the report. More detailed conclusions and recommendations can be found within the individual chapters and are summarized at the end of each chapter.

Although some of the report's conclusions and recommendations revisit issues discussed by previous WSTB committees, the majority of the report concerns the increased emphasis in Cycle II on water quality trends and understanding the causative factors of water quality conditions. It is important to state that the structure of this report, unlike many NRC reports, does not directly follow the committee's statement of task in terms of the order and discussion of major study topics. Rather, the report's structure, and indeed the committee's approach to addressing its charge (see Chapter 8 for an explicit summarization of where and how all four statement-of-task issues are addressed in the report), evolved to reflect discussions with NAWQA staff and especially to be consistent with several iterations of the NAWQA Cycle II Implementation Team (NIT) report *Study-Unit Design Guidelines for Cycle II of the National Water Quality Assessment (NAWQA)*[1] that were provided for the committee's deliberations. As such, different versions of the evolving NIT planning documents (but primarily the November 2000 draft) are cited throughout this report. Indeed, Appendix A of this report includes summaries of all Cycle II goals, themes, and objectives and other important supporting materials extracted from the NIT report. It is also important to state that in presenting such a comprehensive programmatic overview, the committee deemed it necessary to make some general comments about budgets and resources where pertinent to the scope of the proposed changes and additions to the NAWQA program.

At the same time this study was proceeding, a different NRC committee was addressing issues important to the mission of the USGS as a whole. That committee's report, *Future Roles and Opportunities for the U.S. Geological Survey*,[2] was published during the final deliberations for this report. NAWQA itself is a major program within the Water Resources Division of the USGS, and hence this committee's assessment of NAWQA is a microcosm of those larger issues. Also, some recommendations in this report likely go beyond NAWQA's responsibilities and/or capabilities and should be addressed to the broader programs of the USGS. Thus, the committee places its conclusions and recommendations on opportunities to improve NAWQA in the context of that broader report in the hope that it will enhance the utility of both NRC reports.

---

[1] Gilliom, R., et al. 2000. Study-Unit Design Guidelines for Cycle II of the National Water Quality Assessment (NAWQA). U.S. Geological Survey NAWQA Cycle II Implementation Team. Draft for internal review (11/22/2000). Sacramento, Calif.: U.S. Geological Survey.

[2] National Research Council. 2001. Future Roles and Opportunities for the U.S. Geological Survey. Washington, D.C.: National Academy Press.

## INITIAL ASSESSMENT OF NAWQA ACCOMPLISHMENTS

The committee's initial assessment of NAWQA and its representative accomplishments to date finds it a mature and respected national program, with hundreds of publications to its credit and many significant science and policy achievements for the program to build upon. NAWQA has produced not only an unprecedented volume of quality data for use in the scientific community, but also unbiased information that is being used by decision makers, managers, and planners at all government levels. NAWQA has also assumed a vital leadership role, helping to improve environmental monitoring in many agencies from federal to local, both by its example and by technical assistance to others. The use of NAWQA information and the linkages that many other organizations continually seek to make with NAWQA are an illustration of the important void that NAWQA has filled in the national scope of water quality investigations. NAWQA is also to be commended for striving for continual improvement and, to this end, has repeatedly asked for review and critical input from various stakeholders, interest groups, and the NRC.

Despite their accomplishments and impressive legacy of quality reports, NAWQA staff will be increasingly challenged to plan and execute monitoring in Cycle II because of budgetary constraints. For these reasons, NAWQA must continue to review its efficiency and cost-effectiveness; NAWQA staff will have to apply the lessons learned from their first decade of national monitoring and find new ways to operate effectively. With such continued diligence and improvement, NAWQA should be able to meet the challenges and goals that Congress and the nation have asked of it.

## TRANSITION FROM CYCLE I TO CYCLE II: REPRESENTATIVENESS OF STUDY UNITS

Cycle I of NAWQA set out to accomplish its status, trends, and understanding goals with plans to sample 60 study units (reduced to 59 in 1996) that would cover more than half of the nation's land area and account for about 70 percent of the nation's drinking water use.[3] However, because of budgetary constraints, a total of eight study units that were slated for monitoring in 1997-2001 of Cycle I were never initiated (see Figure 1-1 and Table 1-1). Thus, Cycle I of NAWQA included a total of 51 study units and the High Plains Regional Ground Water (HPGW) Study, which was initiated in 1999 (see Chapter 1 for an overview of monitoring in Cycle I). Despite cutbacks, a major effort during Cycle I was the compilation and synthesis of study unit data to make inferences about "regional"

---

[3]It is important to note that water quality conditions are not necessarily studied throughout an entire study unit.

and "national" water quality. Continuing budget constraints have now dictated that the number of study units in Cycle II be reduced from 59 (for planning purposes) to 42, plus the HPGW study (see Figure 2-1 and Table 2-1). Because of this necessary reduction, issues of representativeness and coverage are even more central to Cycle II of the NAWQA program than they were in Cycle I. Assessing the coverage and representativeness of Cycle II is also fundamental to fully address the crosscutting issues of extrapolation and aggregation raised in the committee's statement of task.

Despite the significant reduction in the number of study units to be monitored in Cycle II, the committee feels that the planned study units will still maintain good coverage of the nation's stream and groundwater resources. The primary reason is the commendable and iterative use of hydrologic setting regions, coupled with a linear programming approach (LPA) and expert judgment-based semi-quantitative analysis (SQA) to select the reduced number of study units. As a result of this robust approach, collectively the 42 study units planned for Cycle II account for about 40 percent of the nation's land area, 61 percent of its drinking water use, and 80 percent of the nation's agricultural land.

Nationwide monitoring of lakes, reservoirs, and coastal waters (e.g., estuaries) was specifically excluded from the original NAWQA design. Nevertheless, during Cycle I, certain lakes and reservoirs that provide drinking water and also some lake and reservoir sediments were sampled and analyzed. For this reason and because of limited resources, any Cycle II assessment of lakes and reservoirs will continue to be limited. In addition, although the program has made estimates of chemical fluxes into certain estuaries that are important to other agencies' programs (e.g., the National Oceanic and Atmospheric Administration [NOAA]), NAWQA does not specifically assess coastal waters themselves.

The committee's major recommendations concerning the representativeness of the study unit approach to water quality monitoring and assessment for Cycle II are as follows:

- Maintain sampling in lakes and reservoirs that are important public water supplies, collaborating with other organizations where feasible.
- Clearly state the geographic extent of any inferences that NAWQA scientists may make. This is critical since the NAWQA title implies a nationwide study, but NAWQA is not a statistical sample.
- NAWQA needs to clarify the representativeness of Cycle II study units related to major types of land use that can heavily impact water quality, such as mining, forest products, petrochemical, and related industries.

## ASSESSMENT OF NAWQA CYCLE II GOALS

As noted previously, various iterations of the NIT report *Study-Unit Design Guidelines for Cycle II of the National Water Quality Assessment (NAWQA)* were central to much of the committee's deliberations and are cited frequently throughout this report. Indeed, Chapters 3 to 5 of this report assess the design and implementation strategy for planned investigations in Cycle II study units as related to the continuing program goals of water quality status, trends, and understanding their causes and effects, respectively.

## Status Assessments

The status component of NAWQA is the baseline for all further trend- and understanding-related activities. The various themes, objectives, and contaminants selected for monitoring in Cycle II under the status goal of NAWQA are also fundamental to assess the priority issues selected for national synthesis. As noted previously, the committee was charged to examine the completeness and appropriateness of priority issues already selected for broad investigation under the national synthesis component of the program. The committee supports these existing priority national synthesis topics—pesticides, nutrients, volatile organic compounds, and trace elements—and commends NAWQA for its past work on these important topics. The committee also strongly supports the priority for ecological synthesis that was initiated in the last years of Cycle I. This is also a very important topic to which NAWQA can make a significant contribution.

Reflecting the aforementioned broad shift in focus from gathering occurrence and distribution data to better understanding water quality trends and cause-and-effect relationships, the status goal is slated to receive fewer resources in Cycle II than in Cycle I. Nonetheless, several changes in water quality status assessments have been proposed for Cycle II. To support such status work, three status themes—each with two objectives—were planned for Cycle II.

In Cycle II, NAWQA personnel have proposed to increase the program's focus on the most important previously unsampled stream and groundwater resources—especially those that serve as sources of potable water. The committee concurs with this general strategy, as well as that of focusing on those sources most likely to be impacted by extensive urban and agricultural activities. One important question considered by the committee is whether the significant reduction in the number of study units from Cycle I to Cycle II will have a deleterious effect on achieving these first two status themes and their corresponding objectives. To a large extent, the reduction in study units has been mitigated, as previously discussed, by the robust LPA and SQA methods used to select the reduced suite of Cycle II study units. For this reason, the committee feels that these themes are likely to be accomplished in Cycle II.

Despite the reduced emphasis on conducting status assessments in Cycle II, the committee is pleased that NAWQA proposes to add some new contaminants that have become high national priorities in the last decade to its list of constituents to be monitored in Cycle II. A number of potential candidates have been suggested for monitoring and were assessed by the committee, including methyl mercury, waterborne microbial pathogens, new pesticides, pharmaceutical products, and high-production-volume (HPV) industrial chemicals. Because of the limited funds available, additions must be limited.

The committee provides the following major conclusions and recommendations related to status assessments of water quality in Cycle II:

- Current sampling in lakes and reservoirs that are important public water supply sources should be maintained; other important lake-reservoir public supply sources should be included if resources become available (this might involve a reassessment of which lakes or reservoirs to sample).
- The decision about which additional contaminants to study in Cycle II of NAWQA should be made with direct input from the U.S. Environmental Protection Agency (EPA) and other agencies so that the most important contaminants from a policy-making standpoint can be monitored.
- The NAWQA team should develop a procedure either jointly or with the direct input of EPA and other agencies whereby all contaminants can be evaluated and/or ranked according to a variety of criteria (e.g., known or suspected health or ecological effects) as part of the decision process for inclusion of new contaminants for monitoring.
- All three groups of pesticides proposed for monitoring in Cycle II (important organophosphate insecticides and degradates, several sulfonyl urea herbicides, glyphosate) are appropriate and warranted and should be monitored in Cycle II.
- NAWQA should not add pharmaceuticals or additional HPV industrial chemicals to the list of contaminants to be monitored until reliable sampling protocols and analytical methods can be validated.
- The committee strongly supports the addition of waterborne pathogens and indicator microorganisms to the list of contaminants that will be monitored in Cycle II. However, NAWQA should reconsider its previously proposed microbiological sampling program[4] that includes more detailed sampling (e.g., for both indicator organisms and pathogenic organisms), because waterborne pathogens are of such known importance to human health.
- NAWQA should make sediment a national synthesis topic (i.e., summarize and synthesize findings on sediment, sediment-related pollutants, and habitat impairment from sediment), and provide implications for the nation that can be used by policy makers to maximize the benefits of state and federal conservation resources.
- The USGS should expand its internal expertise in sediment monitoring and interpretation to provide a national leadership role for this important area of

---

[4]Described in Francy, D. S., D. N. Myers, and D. R. Helsel. 2000. Microbiological Monitoring for U.S. Geological Survey National Water-Quality Assessment Program. U.S. Geological Survey Water-Resources Investigations Report 00-4018. Columbus, Ohio: U.S. Geological Survey.

water quality research and work with other local, state, and federal agencies to identify and conduct research on important sediment-related issues.

## Trends and Statistical Support for Understanding

The second goal of the NAWQA program, which receives increased emphasis in Cycle II, is the determination of observed trends (or lack of trends) in water quality. The reliable and early detection of trends is of fundamental value because it can provide information on changes in water quality (especially related to anthropogenic sources) that might be useful for decision making and scientific understanding relating to the management of water quality. If trends are detected successfully in a timely fashion, along with a scientific understanding of the cause of those trends, it may be possible to implement management strategies to help reduce future degradation in water quality. To support this type of causal assessment, Cycle II of the NAWQA program has established three themes and six objectives for the determination of trends in the status of water resources and particularly the effects of urbanization and agricultural practices on water quality.

Regarding the planned studies on the effects of urbanization and agricultural practices on water quality, the proposed assessments will be conducted as space-for-time studies. In this context, "space for time" implies the use of studies on many watersheds and aquifers in space, all characterized by different levels of urbanization and agricultural management practices. Analysis across many watersheds and aquifers in space, at one time, allows for an evaluation of the effects of urbanization and agricultural management practices that normally evolve over time in an individual watershed. Equivalent long-term sampling would require much greater budgetary commitments and would take far more time to get results; hence the proposed space-for-time studies are creative and efficient alternatives. The committee also notes that many of the topics discussed in the assessment of NAWQA's plans for its trend goal for Cycle II are pertinent to addressing the committee's statement of task. This is particularly true for the interrelated issues on extrapolation and aggregation of information at regional and national scales.

The committee finds that the USGS and NAWQA are well positioned to carry out the important work of assessing trends in Cycle II. In this regard, NAQWA has established water quality baselines and monitoring networks in the study units in Cycle I and is operating at time and spatial scales sufficient to establish these relationships. Several major recommendations regarding the ability of the Cycle II NAWQA program to meet these trend themes and related objectives are provided below:

• NAWQA should continue emphasis on an integrated approach to water quality monitoring network design that attempts to coordinate efforts among various local, state, and federal agencies in an effort to make study unit designs as efficient and cost-effective as is possible.

- NAWQA should develop a generalized quantitative approach to optimize the design of water quality monitoring networks that can be tailored to satisfy relevant NAWQA objectives.
- If trend evaluations are to be meaningful, they must account for all relevant statistical complexities to enable meaningful conclusions. Without a proper accounting for the spatial correlation of the water quality time series, resultant conclusions regarding trend assessments are likely be flawed. Future NAWQA research should use methods to account for the spatial correlation of water quality time series.
- NAWQA should continue to emphasize practical and efficient models of the concentration-discharge relationship because such models have applicability to trend assessments as well as many other problems, and it should place greater emphasis on the integration of physical and statistical models because such integrative research should lead to improved and more credible models.
- The NAWQA Cycle II study design has focused on current land-use conditions and their relationship to stream and groundwater attributes. As data analysis proceeds, attention must also be given to urban watershed management practices and the "legacy" effects of past land use.
- To assess water quality impacts resulting from agriculture practices, NAWQA should consider studies related to exemplary state water quality protection initiatives (e.g., California's Pesticide Contamination Prevention Act, Nebraska's Ground Water Management and Protection Act, North Carolina's Nutrient Sensitive Waters).

## Understanding

"Understanding" is the last of the three primary goals for the NAWQA program. Understanding can be gained through the linkage of field studies to the analytical use of models, where observations are compared to a conceptual relationship expressed mathematically. The success of such a model in explaining observations is regarded as a measure of understanding the primary factors or mechanisms involved. Conversely, model development and application require understanding. The studies that provide an understanding of contaminant sources and their transport can be viewed as the raw materials for design and development of water quality models. The status and trends networks for Cycle II provide the basis for understanding studies, and therefore the proposed application of models in Cycle II rests firmly on the knowledge gained from Cycle I.

Models are developed to provide predictions of water quality conditions both spatially and temporally (i.e., through geographic extrapolation or prediction of conditions). Detailed understanding of contaminant sources and their transport to water resources and an ability to predict future conditions are key to development of efficient management and policies to protect the beneficial uses of the nation's water resources, including drinking water and viable ecosystems. As NAWQA progresses into Cycle II with an increased emphasis on its understanding goal, the importance of model application, as recommended by previous NRC committees, should not be underestimated. Understanding and prediction, embodied in water quality models, are the cornerstones of water resources management for the future.

Several types of models are proposed for application in Cycle II. The committee presents an overview of models, including those categorized as conceptual, mass balance, statistical regression, process based (mechanistic), and hybrids of these. The systematic arrangement of these models in a hierarchy of spatial and temporal scales should be linked to and will advance the development of the Modular Modeling System (see Chapter 5). When more complex models are applied, the focus should be on those that can be parameterized with data. The quantification of uncertainty is important in providing perspective for the interpretation of model results, and this will require some specialized focus within NAWQA.

The basic NIT plan is for "targeted studies" to focus on a limited set of the most important water quality topics and for linkage of these field studies with models and other parts of the Cycle II design. Each specific targeted study will be designed and executed by "topical teams," composed of one or more Cycle II study units, and they will be assisted by a single, national Hydrologic Systems Team (HST) that will provide modeling guidance to the study units. These targeted studies will focus on the major factors that govern water quality. The contaminants studied in Cycle I (e.g., pesticides, nutrients, volatile organic compounds) and the new drinking water source status assessments planned for Cycle II are expected to provide the foundation for targeted study design wherever possible.

The six themes with 17 objectives that describe the understanding goal for Cycle II were developed (in the preliminary planning) by categorization of the wide variety of scientific studies proposed by the Cycle II study units. This is the first step in determining where commonalties or synergies between studies exist on a regional or national scale. A greater distillation of this collection of ideas is needed since it does not appear that the objectives have been sufficiently refined and focused. However, the final section of the NIT guidance presents an example for targeted study development by an Agricultural Chemical Source and Transport Team (ASTT) and includes a set of more clearly defined hypotheses for study. It also may be useful to relate the proposed studies and their components to the conceptual "source-transport-receptor" (STR) model presented in the NIT guidance. The STR model could be used to organize a wide range of studies from across the nation to show how they relate to each other and where information may be lacking. It is important that the targeted studies are interpreted in the context of NAWQA's themes and related objectives for Cycle II so that they may be evaluated in terms of geographic applicability and national priority.

At this important juncture in the development of NAWQA, the committee concludes that the USGS has several major opportunities to advance scientific

EXECUTIVE SUMMARY

understanding of factors that affect water quality conditions. However, the committee is concerned whether or not sufficient staff, resources, and expertise are allocated to ensure that modeling efforts and targeted studies can be developed and implemented adequately. While the NIT report recognizes the resource problem, its resolution will ultimately govern what can be accomplished in Cycle II. Given these general conclusions from the review of NAWQA plans for the understanding goal in Cycle II, the committee provides the following major recommendations:

- Conducting mass balances on constituents of concern is a worthy effort in Cycle II. It is strongly recommended that at least a conceptual mass balance be developed for the nonpoint source pollutants studied in each study unit. NAWQA and the USGS should focus on simple, parsimonious process models (i.e., models that are not overparameterized), where parameter estimation and mechanistic expressions can relate to available data.
- The HST should ensure some uniformity (and/or compatibility) in the use of models and software packages in Cycle II to enable national comparisons and aggregation of data and results. Furthermore, to test research hypotheses, the HST must also address error terms and uncertainty in model selection and application. In this regard, all future NAWQA studies should attempt to evaluate uncertainty associated with all aspects of its water quality modeling evaluations, and the HST should develop a guidance document on this topic.
- NAWQA should establish a statistical model support team (perhaps as a subgroup within the HST) to provide guidance on the selection and application of the increasingly important modeling tools. This group can build on the USGS's historical strength and experience in regionalization and extrapolation of streamflow. Also the group should explore and advise on various issues, including more parsimonious and adaptive models, new techniques (e.g., Bayesian), and previously discussed issues such as uncertainty assessment.
- The process of selecting topics for targeted studies to meet the understanding goal in Cycle II has not yet been clearly defined. The example of the ASTT, however, was well done, presenting clearly defined hypotheses and a focus lacking in other areas. The committee recommends that NAWQA build on this example to develop and refine its targeted study approach.
- NAWQA must resolve staffing, expertise, and resource issues if the objectives of the understanding goal are to be addressed successfully in Cycle II. It is not clear whether sufficient staff expertise or resources are available for model application and targeted study implementation.
- For the identification of contaminant sources, when feasible, monitoring programs should employ highly specific or supplementary analyses that permit distinction between different sources of the same contaminant (e.g., ribotyping of microbiological specimens, isotopic analyses of some chemicals).
- The hyporheic zone, or groundwater-surface water (GW-SW) ecotone, is

an important component in understanding surface water quality and near-surface groundwater quality. All appropriate Cycle II study units should endeavor to (1) design a process-based approach to characterize GW-SW interactions and their effects on water quality and (2) use process-based models that can include GW-SW interaction components to delineate the spatial and temporal variations in GW-SW interchange and the concomitant water quality changes. This may mean seeking out collaborators and cooperators to find the needed expertise and study sites, and this should be encouraged as well.

• Habitat degradation is often argued to be the most significant cause of ecological impairment. Thus, NAWQA should reconsider including studies on the impact of alterations in hydrologic regime and sediment transport in its later-phase Cycle II plans.

• Although there may not be a specific objective relating to exotic species, data already collected and being analyzed by NAWQA will be valuable for examining questions about their impacts. NAWQA should find ways to encourage this synthesis, perhaps by developing cooperative arrangements within the U.S. Fish and Wildlife Service's Invasive Species Program.

• The USGS's NAWQA program is in excellent position to make an important contribution to the debate on which biological indices provide the most meaningful assessments of water quality conditions based on results from Cycle I studies. This should be a top priority of the Ecological Synthesis Team. In addition to evaluating different indices, it is critical that the Ecological Synthesis Team explore quantitative relationships and potential threshold responses between biotic indices and other measures of water quality.

## COMMUNICATING NAWQA DATA AND INFORMATION TO USERS

The NAWQA program is first and foremost a provider of information to parties interested in water quality. Although most of the NAWQA budget and effort are devoted to data collection and interpretation, it is the reporting of the program's findings that is most critical for its widespread use and ultimate success. In this regard, the NAQWA program has generated an impressive amount of information since its inception and has kept the public reasonably well informed of its plans and findings. Both the national synthesis teams and the individual study units are providing useful information on all facets of the program, including sampling design, implementation issues, results, and interpretations. Information is being conveyed through several types of written reports, journal articles, professional papers, digital products, the NAWQA Internet site, and an on-line Data Warehouse. The USGS is taking pains to provide guidelines to NAWQA staff for producing effective reports. Similar guidance would improve the quality of information provided by national synthesis teams and study units through the Internet. NAWQA has recently started using Internet-based "briefing rooms" to convey findings that bear on important water quality issues.

## EXECUTIVE SUMMARY

This is an effective approach for providing relevant information to policy makers and those interested in water quality issues.

Many states are using NAWQA data and findings in developing their resource management programs. This is a strong indication that NAWQA is providing valuable information to those managing water resources. However, the committee finds that NAWQA can improve on the ways it conveys information to policy makers, resource managers, and the public:

- The USGS and NAWQA need to clearly report in summary fact sheets and on the NAWQA home page changes in the scope of the program and future plans in a timely manner.
- NAWQA should consider the formation of a distinct information office that would provide additional resources to the important task of timely and efficient information dissemination. This office could also explore innovative strategies for getting information to policy makers, resource managers, and the public.
- NAWQA should be able answer basic but very important questions about the nation's water quality such as, Has it improved since the passage of the Clean Water Act? While NAWQA's ongoing national synthesis work represents an important first step in addressing this type of question, explicit answers could be particularly valuable to Congress, policy makers, resource managers, and the public. Even a prospective discussion of how NAWQA can or will eventually answer such questions in Cycle II would be useful.
- The NAWQA Leadership Team should continue to work with national synthesis teams and individual study units to maintain and improve the quality of written reports, to ensure that the needs of policy makers are met, and to improve the content and consistency of NAWQA Web sites.
- NAWQA should expand the use of Internet-based briefing rooms to convey important information, rather than relying exclusively on electronic versions of USGS documents.
- NAWQA should encourage its staff to continue to publish NAWQA findings in refereed professional journals, where they will receive review from other scientists and broader exposure to the scientific community.

## COOPERATION AND COORDINATION

The national scope of NAWQA and its potential to provide a nationwide perspective on the status, trends, and understanding of factors that affect water quality have made it a focal point within the USGS. Indeed, many local, state, and even federal agencies and organizations, many of which have not worked with the USGS in the past, now regularly promote the use of NAWQA data and information (though sometimes without a full understanding of their inherent limitations and availability). However, the increased use and visibility of NAWQA data and information often occur in conjunction with attempts to influence the design or to cooperate to broaden NAWQA's coverage.

With a program of such scope, cooperation, coordination, and real collaboration with external agencies and organizations (e.g., related to budgets and staffing) should be priorities to optimize the massive data collection, as well as data use and interpretation efforts. Indeed, every chapter of this report includes examples of such cooperative efforts. Through these examples, the committee hopes to address the various cooperation and coordination concerns and issues raised at committee meetings or during interviews and meetings with USGS and NAWQA personnel and others with a vested interest in NAWQA data and information. These examples address both benefits and problems for the program, illustrate typical management challenges, and may identify some new or expanded opportunities for cooperation and collaboration. In the committee's view, NAWQA program staff have done an excellent job of establishing cooperative relationships within USGS and with external programs. These efforts have strengthened NAWQA and have improved the visibility and viability of the USGS as a whole. Cooperation has costs, however, particularly in staff time to keep effective communications in place. Also, while cooperative efforts are valued, NAWQA cannot be and should not attempt to be all things to all people—or agencies.

In this regard, other agencies that want to utilize NAWQA or to coordinate programs with NAWQA also have a responsibility to collaborate fully with the program (i.e., to give, not just take). As large as NAWQA is, its program resources are often too constrained to fully meet its national goals. Continued flat budgets or budget cuts will not allow NAWQA to meet its goals or to provide the information that Congress and other agencies desire without continued design changes and cutbacks. Certainly, programmatic and design changes that can improve the efficiency and cost-effectiveness of NAWQA are always warranted and should continue to be developed, but providing a national perspective on the nation's water quality requires adequate support. While Congress must recognize this, other agencies also have to contribute their due support, with staffing, fiscal, or in-kind support where possible, to help cover their unique needs and requests. The USGS must also try to ensure reasonable overhead to keep transaction costs affordable for cooperators.

The committee finds that these cooperative efforts have strengthened NAWQA and have improved the visibility and viability of the USGS as a whole. However, in the area of program coordination, the committee offers the following major recommendations:

• NAWQA must stay firm in its design to meet its national goals, and it should not change critical design plans to meet the diverse needs of the many federal, state, and local agencies. Thus, NAWQA must maintain its careful balancing act to uphold its design principles that draw other agencies to NAWQA, while finding ways to collaborate that build and improve NAWQA. Perhaps

EXECUTIVE SUMMARY

more importantly, such collaboration should strive to improve and strengthen other water-related programs to enhance the total knowledge of the nation's water resources.
- For the continued success of NAWQA, the USGS must continue to manage the "creative tension" of the joint efforts among its internal programs. Based on the committee's investigations and deliberations during this study, several potential areas of stress require continuing dialogue: (1) the management problems created by the fluctuation in resources and staffing that occurs between high-intensity and low-intensity monitoring phases of NAWQA should be addressed; (2) district-level staff must fully appreciate the national goals of NAWQA, and likewise NAWQA personnel should listen to implementation concerns from their local expertise; (3) the proposed coordinating committee for NAWQA and the USGS's National Stream Quality Accounting Network (NASQAN) should be implemented; and (4) NAWQA must continue to recognize and share full and proper credit with its partners.
- The local study unit and national liaison committees should certainly be continued in Cycle II. NAWQA should consider providing training as well as further information sharing (lessons learned) for study unit teams to make the local liaison efforts more beneficial for USGS and local users.
- The USGS should continue to foster the use of the good science produced by NAWQA with its state cooperators. Of particular note, USGS scientists should support opportunities to use NAWQA analyses and the SPARROW (Spatially Referenced Regressions on Watershed Attributes) watershed-scale model in the development of total maximum daily loads (TMDLs). TMDLs, however, are the states' responsibility. Thus, NAWQA resources and scientists should not be diverted to working on TMDLs beyond the data and technical assistance that they can readily provide to the states.

## THE FUTURE OF NAWQA

The title of this report and the committee's statement of task clearly indicate issues that will affect the NAWQA program and staff in the future, but they do not fully convey the potential importance of this programmatic review. In the last decade-and-a-half, NAWQA has progressed from a sound concept to a mature program of exemplary quality and importance. NAWQA has led the way to begin the critical, sound scientific assessment of the quality of the nation's waters. Because of this initial success, NAWQA now carries with it high expectations from many other federal, state, and local agencies, as well as policy makers and legislators. At this juncture, NAWQA is in a critical period as it transitions from Cycle I to Cycle II and as many aspects of the promise of NAWQA must come to fruition. In this regard, the committee hopes that the timing of this review and report can contribute to the future of NAWQA—a future that has clearly become of widespread importance to the nation.

The design and management of NAWQA and its past, present, and future success is an ongoing struggle for program balance, and this is reflected in the committee's conclusions and recommendations. NAWQA must strive to find the appropriate balance of efforts and resources between its three primary goals of status, trends, and understanding as it enters its second decade of nationwide monitoring. The committee fully expects that NAWQA will continue to exhibit foresight and take a lead in studying emerging water quality issues and contaminants and will avoid expending unwarranted resources on a "contaminant-of-the-day" approach. As another exemplary balance issue, the committee notes the very important contributions that NAWQA can and should make in biological assessments and ecological synthesis in surface waters, yet the committee also strongly recommends that NAWQA not embark on an ecotoxicology program. Furthermore, although NAWQA must strive to be responsive to water quality policy and regulatory needs, it cannot be driven or controlled by these needs—thus epitomizing the struggle of doing "good science" in the public policy arena. In this regard, the committee commends NAWQA for doing an excellent job of balancing good science with policy needs in the face of flat budgets. However, NAWQA supporters, users, policy makers, and Congress itself should be made aware of the fine balancing act that this requires, and they should be supportive of the dilemma that it creates for operating such a program. NAWQA must also balance its work within the context of the agency that provides its charter, the USGS.

This committee, and nearly all users of NAWQA with which it has interacted, recommend that NAWQA do more, not less, even amidst the obvious resource constraints faced by the program. NAWQA's resources have not grown to keep pace with annual inflation, and it has had to redesign significantly for Cycle II. Although NAWQA has done an exemplary job of downsizing to 42 planned study units for Cycle II, it cannot continue to downsize and still be considered a *national* water quality assessment. Though it could certainly be redesigned, this would likely undo the basis for assessment of trends and would waste a decade or more of effort.

To address long-term trends in water quality across the nation, one must recognize the concurrent need for long-term support to allow for consistency in the data gathering and analysis efforts. As discussed throughout this report, the future success of NAWQA is a struggle for balance of resources and scientific endeavors in a water policy environment. Current and future demands already exceed the capacity of NAWQA, but it is hoped that policy makers, politicians, and program managers can strike the necessary balance that will allow NAWQA to continue to provide important water quality data and information for the nation.

# 1

# Introduction

Since its creation by an act of Congress in 1879, the U.S. Geological Survey (USGS) has devoted a substantial portion of its personnel and budget to studying the nation's water resources. As a result, this agency of the Department of the Interior has been looked to historically as the primary national source of information on water quantity and quality in the United States, including both surface waters (e.g., lakes, rivers, estuaries) and groundwater. Subsequent to the increased national interest in environmental issues and the passage of extensive environmental legislation in the 1970s, the USGS role in providing data and information on water quality has become even more pivotal and more highly scrutinized. During congressional testimony in the mid-1980s, scientists at USGS realized that their ability to understand the quality of the nation's water as a whole was limited by sporadic and variable monitoring networks across the country that did not allow national-scale data analysis. Thus, the National Water Quality Assessment (NAWQA) Program was created in 1985 to better assess historical, current, and future water quality conditions and trends in representative river basins and aquifers across the United States using highly consistent methods.

A primary objective of the NAWQA program is to describe relationships between natural factors, human activities, and water quality conditions and to define those factors that most affect water quality in different parts of the nation. More specifically, three goals drive NAWQA's design and development: (1) *status*—to provide a nationally consistent description of the current water quality conditions for a large part of the nation's water resources; (2) *trends*—to define long-term trends (or lack of trends) in water quality; and (3) *understanding*—to identify, describe, and explain (to the extent possible), the major factors that affect (and cause) observed water quality conditions and trends. Although the exact wording of the goals has been refined over time, these three goals are the organizing themes for NAWQA's past, present, and future.

During its first decade of monitoring (called Cycle I; see more below), which spanned from 1991 until 2001, NAWQA concentrated primarily on gathering comparable information on water resources (i.e., status assessments) in dozens of geographic areas, called study units, that include major river basins and/or aquifers nationwide. Perhaps the most important facet of the program is that the similar design of each investigation and the use of standardized methods make comparisons among disparate study units possible. Combining, comparing, and analyzing data from individual study units has led to regional and national assessments of water quality, collectively referred to as "national synthesis." More specifically, national synthesis also includes information from other programs, agencies, and researchers to produce regional and national assessments for priority water quality issues. At first, national synthesis in Cycle I concentrated on two issues of national priority—the occurrence of (1) nutrients and (2) pesticides in streams and groundwater. These topics were ranked among the highest in importance because of widespread environmental and public health concerns and because the information necessary for a national assessment of these contaminants was incomplete. Later topics addressed by the national synthesis component of NAWQA included volatile organic compounds, trace elements, and ecological synthesis (in that order).

The National Research Council's (NRC's) Water Science and Technology Board (WSTB) has provided advice to NAWQA four separate times in the past as the NAWQA program has evolved from an unfunded concept in 1985 to a relatively mature and tested program in 2001. In a letter report from Walter Lynn, then the WSTB chair, to former USGS director Dallas Peck in October 1985, the WSTB heartily endorsed the concept of the original NAWQA program. WSTB members raised certain issues for USGS consideration during NAWQA program development. First, in response to the strong focus of data collection on physical and chemical parameters, the WSTB stressed the importance of including biological parameters in the program. It felt that both short-term and long-term impacts of water quality on human and ecological health would be best assessed via biological assessment. The board promoted increasing the level of interaction between the USGS and other federal and state agencies to maximize the utility of NAWQA data. Finally, the difficulties inherent in measuring groundwater parameters were acknowledged and a pilot program on groundwater quality assessment was suggested.

Shortly after this initial communication, the WSTB convened a colloquium (NRC, 1986) to specifically address what the necessary elements should be for a national water quality monitoring and assessment program. Colloquium participants raised new issues for consideration, such as which particular chemical constituents should be chosen for measurement, whether and how to interface with state monitoring programs, and how the program could reflect surface water-groundwater interactions. Keynote speaker William Ackermann (deceased; then of the University of Illinois) warned that a national monitoring network, even if designed optimally, cannot be all things to all users. However, it should strive to meet the highest technological standards, and its data and information should be made widely available. Colloquium participants supported the idea that to initiate the program it should be operated by a single federal agency continuously for at least five years.

NAWQA was initiated in 1987 via a pilot program that focused on seven study units, all of which were either surface water or groundwater dominated. Later that year, the WSTB was requested to undertake a two-year evaluation of the pilot program and make recommendations for full-scale implementation of NAWQA. The committee that was formed to conduct the study unanimously agreed that that there was a genuine need for a national water quality program and that the USGS was ideally situated to manage such a program. After evaluating five early NAWQA reports, visiting pilot study locations, and interviewing users of NAWQA products, the NRC committee released its final report, *A Review of the USGS National Water Quality Assessment Pilot Program*, in 1990.

That report recommended that during full-scale implementation, some study units should be selected in which both surface water and groundwater are jointly investigated to better understand such integrated systems. Regarding possible program deficiencies, the committee questioned the decision not to include lakes and estuaries in the monitoring program and recommended that they be included whenever possible. To conserve limited resources, the pilot program was designed to rotate between study units on a regular basis, a strategy with which the committee agreed but felt should be reviewed after completion of the first phase of the program. The committee was generally pleased with the choice of chemical and physical constituents but, as in earlier WSTB communications, felt that the chosen biological parameters needed refinement. The report noted that the NAWQA pilot program at that time lacked a quantitative framework for evaluating cause-and-effect relationships between pollution sources and nearby waters, particularly at large scales. It recommended the development of mathematical modeling to assess cause-and-effect relationships and suggested directing greater resources toward these efforts.

The 1990 report states that "[NAWQA] program documentation is vague on how the information from the study units will be scaled up to the national level so that a national synthesis can be accomplished." Shortly after commencement of the full-scale program in 1991, the USGS sought the WSTB's advice on that very topic. The resulting report *National Water Quality Assessment Program: The Challenge of National Synthesis* (NRC, 1994) discusses the issues that are critical for successful national synthesis, particularly the need for process-oriented research and associated mathematical modeling. The report does not promote the use of specific models, although it notes that mass-balance models should serve as the starting point for conducting national synthesis. It also suggests that a special program be created within NAWQA to develop basin-scale analysis tools necessary for national synthesis. Finally, because of the presumed importance of land-use impacts on water quality, the report recommends a strong effort to compile and make available land-use and land-cover data in a standardized format. Such efforts are readily apparent in many subsequent NAWQA publications.

Nearing completion of the first full cycle of data collection, USGS scientists are planning for the program's future and have again requested the input of the NRC to help shape NAWQA activities during the program's second decade of monitoring, known as Cycle II. An overarching concern of the USGS is to maximize program usefulness for a variety of stakeholders, given limited resources and the inherent limitations of the study unit approach. Public interest in water quality information continues to increase, and the potential applicability of NAWQA data and analyses goes well beyond a simple assessment of the nation's overall water quality. Indeed, information from the NAWQA program is becoming integral to research, monitoring, and regulatory activities at the local, state, and regional levels.

In addition to conducting an initial assessment of the general accomplishments of the NAWQA program to date and making suggestions for the program's overall improvement, the USGS asked the committee to provide comment on four particular areas of NAWQA: (1) suggest methodologies to improve understanding of the causative factors affecting water quality conditions; (2) assess whether information produced in the program can be extrapolated to allow inferences about water quality conditions in areas not studied intensely under NAWQA; (3) examine current priority issues (e.g., pesticides, nutrients, volatile organic compounds, trace elements) selected for broad investigation under NAWQA for completeness, and suggest other pollutants (e.g., microbial pathogens, sediments) that might also be studied on a national basis; and (4) make recommendations on the aggregation and presentation of information generated at the study unit scale so that it is meaningful at the regional and national levels.

This report presents the WSTB's most recent effort to advise the USGS on the NAWQA program as it completes its first cycle of intensive data collection, which was focused on water quality conditions in the United States. Although some of the report's conclusions and recommendations revisit the issues discussed by previous WSTB committees, the majority of the report concerns the increased emphasis in Cycle II on defining long-term water quality trends and understanding the causative factors of water quality conditions. It is important to state that the structure of this report, unlike many NRC reports, does not directly follow the committee's statement of task in terms of the order and discussion of major study topics. Rather, the report's structure, and indeed the committee's approach to addressing its charge, evolved to reflect discussions with NAWQA staff and especially to be consistent with several iterations of the NAWQA Cycle II Implementation Team (NIT) report *Study-Unit Design Guidelines for Cycle II of the National Water Quality Assessment (NAWQA)* (Gilliom et al., 2000b) that were provided for the committee's deliberations (see Chapter 8 for further information). It is also important to state that in presenting such a comprehensive programmatic overview, the committee felt it was necessary to make some general comments about budgets and resources where pertinent to the scope of the proposed changes and additions to the NAWQA program.

The aforementioned NIT report describes the design and implementation strategy for Cycle II investigations in NAWQA study units regarding the major change in emphasis among the three NAWQA goals along with their corresponding themes and objectives. As such, it is cited frequently throughout this report. Indeed, the content, conclusions, and recommendations of Chapters 3 to 5 (see more below) are targeted to a careful review of the NIT report and subsequent deliberations by the committee. Chapters 3 to 5 differ further from the remaining chapters in that they contain sectional summaries and recommendations whereas the others do not. (However, with the exception of this introductory chapter, which has no recommendations, all chapters end with a comprehensive conclusion and recommendation section.) This format was deemed appropriate and necessary given the large number of distinct topics that are addressed in the three central chapters. While many of the committee's recommendations were influenced by or even directly result from a careful review of this important NIT report, the committee did not directly comment on every proposed Cycle II theme and objective since many were fundamental to NAWQA operations and did not warrant discussion or comment. (The committee directs the interested reader to Appendix A of this report, which includes summaries of all Cycle II goals, themes, and objectives, and other important supporting materials that are taken directly from the NIT report.)

At the same time that this study was proceeding, a different NRC committee was addressing issues important to the mission of the USGS as a whole. That committee's report *Future Roles and Opportunities for the U.S. Geological Survey* (NRC, 2001) was published during the final deliberations for this report. Obviously, future opportunities for NAWQA should comprise a subset of opportunities for the USGS. Thus, in Chapter 8, the committee also places its conclusions and recommendations for NAWQA within the broader perspective of that NRC (2001) report to the USGS.

This introductory chapter provides a brief overview and some representative results and accomplishments of NAWQA's first decade of national water quality monitoring and introduces the forthcoming second cycle. Chapter 2 describes the transition and design changes from Cycle I to Cycle II in terms of coverage, representativeness, and other related issues. Chapters 3 to 5 assess the design and implementation goals, themes, and objectives of the USGS for Cycle II of NAWQA as related to water quality status, trends, and understanding of their cause and effects, respectively. Chapter 6 provides a review and recommends ways to improve the communication of NAWQA data and findings to a wide variety of users. Chapter 7 discusses opportunities for increased cooperation and coordination between NAWQA personnel and local, state, and federal agencies that are concerned with water quality. Lastly, Chapter 8 summarizes the report's key conclusions and recommendations as related to the committee's statement-of-task issues and, more broadly, to the future of the NAWQA program.

It is also important to state that the committee's four specific tasks related to its charge are fundamentally interwoven throughout the planned Cycle II goals, themes, and objectives. As such, they are addressed in many places throughout the report, where appropriate, in the context of NAWQA's Cycle II design and implementation planning framework. Issue 1, to "recommend methods for the improved understanding of the causative factors affecting water quality conditions," is addressed primarily in Chapter 5, but pertinent issues are also discussed in Chapters 3 and 4. Issue 2, to "determine whether information produced in the program can be extrapolated so as to allow inferences about water quality conditions in areas not studied intensely in NAWQA," is addressed in Chapter 2 in terms of the representativeness of the reduced study unit framework; in Chapters 3 and 4 in discussing the adequacy of monitoring and monitoring networks; and further in Chapter 5 in terms of extrapolation and forecasting. Issue 3, to "assess the completeness and appropriateness of priority issues . . . selected for broad investigation under the national synthesis component of the program," is addressed primarily in Chapter 3, although Chapters 4 and 5 also comment on related topics such as the effects of urbanization, changing agricultural management practices, and the assessment of stream ecology. Issue 4, to "describe how information generated at the study unit scale can be aggregated and presented so as to be meaningful at the regional and national levels," is addressed at various levels in Chapter 2, again related to the discussion of representativeness and the language of national synthesis; in Chapter 5, as related to extrapolation; and in Chapter 6, as related to meaningfully communicating information to regional and national policy makers. Finally, Chapter 8 explicitly summarizes where and how all four statement-of-task issues are addressed throughout this report.

In summary, throughout its deliberations, the committee maintained that its fundamental and overriding charge was clearly "to provide guidance on opportunities to improve the NAWQA program"—not simply to address four particular issues interwoven in the Cycle I and planned Cycle II design of the program. In this regard, many of the conclusions and recommendations of this report go beyond those implied by a strict reading of the issues raised in the statement of task. For these reasons, the report's organization is essentially a blend of a review of the NAWQA implementation plans for Cycle II and the committee's statement of task.

## OVERVIEW OF CYCLE I OF NAWQA

As noted previously, the goals of the NAWQA program are to assess the status of and trends in the quality of the nation's surface and groundwater

# INTRODUCTION

resources and to link the status and trends with an understanding of the natural and human factors that affect water quality (Gilliom et al., 1995; Hirsch et al., 1988). To this end, NAWQA is completing its first decade of nationwide study (1991 to 2001; Cycle I) and is about to embark on its second decade of study (Cycle II). While this report provides conclusions and recommendations primarily cogent to Cycle II of NAWQA, an overview of the design, implementation, and representative results of its first cycle is prudent because much of the basic design characteristics will remain in place for its second decade. Indeed, NAWQA has been designed to build and expand on the broad base of knowledge gained in its first decade to support the Cycle II emphasis on understanding and explaining the processes controlling water quality (Mallard et al., 1999). This improved understanding will help make NAWQA even more relevant and useful to resource managers, policy makers, and regulatory agencies throughout the United States.

The NAWQA program seeks to balance the unique assessment requirements of individual hydrological systems with those of a nationally consistent design structure that incorporates a multiscale, interdisciplinary approach for groundwater and surface water (Gilliom et al., 1995, 1998; Miller and Wilber, 1999). The program integrates diverse information about water quality at a wide range of temporal and spatial scales and focuses on water quality conditions that affect large areas of the United States. As noted previously, nutrients and pesticides have been the initial focus of much work since the program's inception (Gilliom et al., 1995; USGS, 1999).

The basic building blocks of the NAWQA program in Cycle I continue to be the study unit investigations of major hydrologic basins located throughout the nation. While the original design of NAWQA called for 60 study units, their number in Cycle I was reduced to 59 in 1996 by combining two adjoining study units in New England that were determined to have similar hydrologic and land-use characteristics. At the time this report was written, a total of 51 Cycle I study unit assessments had been initiated since 1991. Of these, 36 assessment reports were completed, with another 15 due to be completed later in 2001. In this regard, eight study unit investigations slated to begin in fiscal year 1997 were never initiated because of budgetary constraints (see Figure 1-1, Table 1-1, and Chapter 2 for further information). In addition, initiated during Cycle I (1999) of NAWQA and continuing into Cycle II, the USGS has a six-year monitoring program to evaluate groundwater quality in the High Plains aquifer system that underlies 174,000 square miles in parts of eight states (Colorado, Kansas, Nebraska, New Mexico, Oklahoma, South Dakota, Texas, and Wyoming; see Figure 1-1). This ongoing NAWQA investigation is called the High Plains Regional Ground Water (HPGW) Study and covers much of the geographic area of the study units that were not conducted because of budget constraints.

Chapter 2 of this report includes a detailed discussion and assessment of the number of study unit investigations planned for Cycle II. Collectively, the

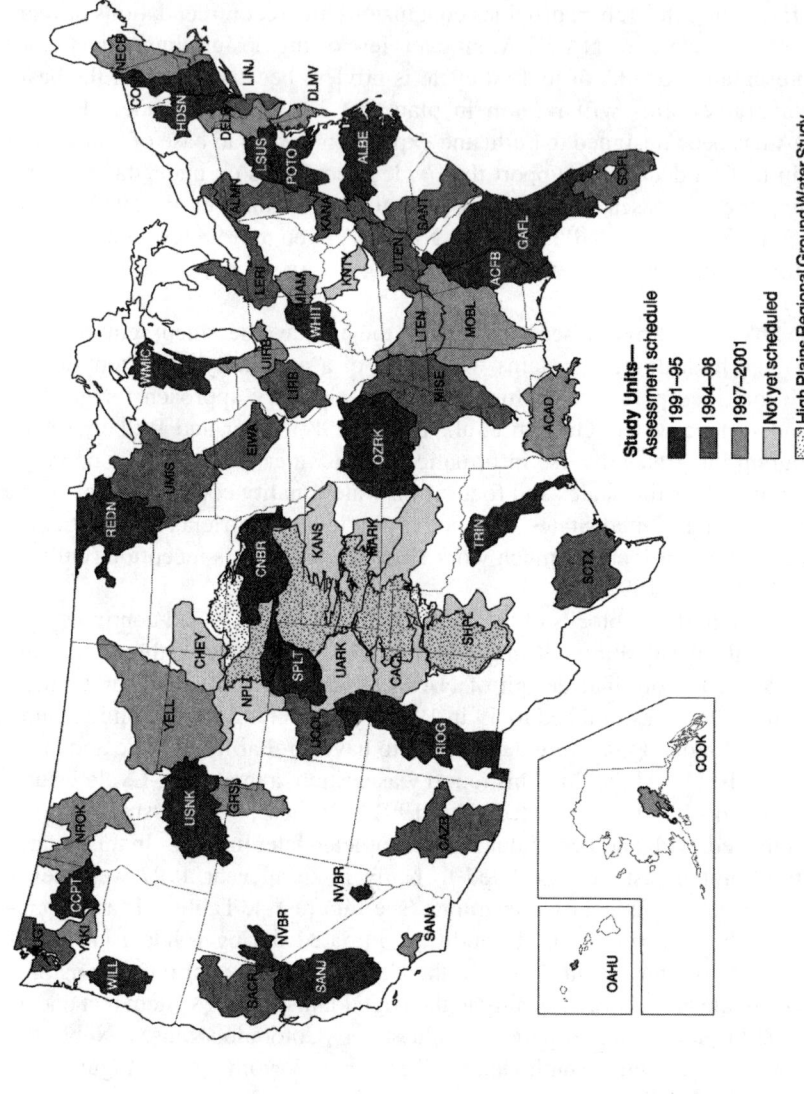

FIGURE 1-1 Monitored and planned Cycle I study units (1991-2001). SOURCE: Gilliom et al., 2001.

# TABLE 1-1 Monitored and Planned Cycle I NAWQA Study Units

| Abbreviation | Name | States in Study Unit | Starting Year |
|---|---|---|---|
| ACAD | Acadian-Pontchartrain Drainages | LA, MS | 1997 |
| ACFB | Apalachicola-Chattahoochee-Flint River Basins | AL, FL, GA | 1991 |
| ALBE | Albemarle-Pamlico Drainages | NC, VA | 1991 |
| ALMN | Allegheny-Monongahela River Basins | MD, NY, PA, WV | 1994 |
| CACI | Canadian-Cimarron River Basins | CO, KS, NM, OK, TX | None |
| CAZB | Central Arizona Basins | AZ | 1994 |
| CCPT | Central Columbia Plateau | ID, WA | 1991 |
| CHEY | Cheyenne-Belle Fourche River Basins | NE, SD, WY | None |
| CNBR | Central Nebraska Basins | NE | 1991 |
| CONN | Connecticut, Housatonic, and Thames River Basins | CT, MA, NH, NY, RI, VT | 1991 |
| COOK | Cook Inlet Basin | AK | 1997 |
| DELR | Delaware River Basin | NY, NJ, PA | 1997 |
| DLMV | Delmarva Peninsula | DE, MD, VA | 1997 |
| EIWA | Eastern Iowa Basins | IA, MN | 1994 |
| GAFL | Georgia-Florida Coastal Plain Drainages | FL, GA | 1991 |
| GRSL | Great Salt Lake Basins | ID, UT, WY | 1997 |
| HDSN | Hudson River Basin | CT, MA, NJ, NY, VT | 1991 |
| KANA | Kanawha-New River Basin | NC, VA, WV | 1994 |
| KANS | Kansas River Basin | CO, KS, NE | None |
| KNTY | Kentucky River Basin | KY | None |
| LERI | Lake Erie-Lake St. Clair Drainages | IN, MI, OH, PA, NY | 1994 |
| LINJ | Long Island-New Jersey Coastal Drainages | NJ, NY | 1994 |
| LIRB | Lower Illinois River Basin | IL | 1994 |
| LSUS | Lower Susquehanna River Basin | MD, PA | 1991 |
| LTEN | Lower Tennessee River Basin | AL, GA, TN | 1997 |
| MARK | Middle Arkansas River Basin | CO, KS, OK | None |
| MIAM | Great and Little Miami River Basins | IN, OH | 1997 |
| MISE | Mississippi Embayment | AR, KY, LA, MO, MS, TN | 1994 |
| MOBL | Mobile River Basin | AL, GA, MS | 1997 |
| NECB | New England Coastal Basins | MA, ME, NH, RI | 1997 |
| NPLT | North Platte River Basin | CO, NE, WY | None |
| NROK | Northern Rockies Intermontane Basins | ID, MT, WA | 1997 |
| NVBR | Nevada Basin and Range | CA, NV | 1991 |
| OAHU | Oahu | HI | 1997 |
| OZRK | Ozark Plateaus | AR, KS, MO, OK | 1991 |
| POTO | Potomac River Basin | DC, MD, PA, VA, WV | 1991 |
| PUGT | Puget Sound Basin | WA | 1994 |
| REDN | Red River of the North Basin | MN, ND, SD | 1991 |
| RIOG | Rio Grande Valley | CO, NM, TX | 1991 |
| SACR | Sacramento River Basin | CA | 1994 |

*continues*

TABLE 1-1 Continued

| Abbreviation | Name | States in Study Unit | Starting Year |
|---|---|---|---|
| SANA | Santa Ana Basin | CA | 1997 |
| SANJ | San Joaquin-Tulare River Basins | CA | 1991 |
| SANT | Santee River Basin and Coastal Drainages | NC, SC | 1994 |
| SCTX | South-Central Texas | TX | 1994 |
| SHPL | Southern High Plains | NM, TX | None |
| SOFL | Southern Florida Drainages | FL | 1994 |
| SPLT | South Platte River Basin | CO, NE, WY | 1991 |
| TRIN | Trinity River Basin | TX | 1991 |
| UARK | Upper Arkansas River Basin | CO, NM | None |
| UCOL | Upper Colorado River Basin | CO, UT | 1994 |
| UIRB | Upper Illinois River Basin | IL, IN, WI | 1997 |
| UMIS | Upper Mississippi River Basin | MN, MI | 1994 |
| USNK | Upper Snake River Basin | ID, MT, NV, UT, WY | 1991 |
| UTEN | Upper Tennessee River Basin | GA, KY, NC, SC, TN, VA | 1994 |
| WHIT | White River Basin | IN | 1991 |
| WILL | Willamette Basin | OR | 1991 |
| WMIC | Western Lake Michigan Drainages | MI, WI | 1991 |
| YAKI | Yakima River Basin | WA | 1997 |
| YELL | Yellowstone River Basin | MT, ND, WY | 1997 |

59 Cycle I study units would have accounted for 60 percent to 70 percent of national water use and of the population served by public water supplies (Gilliom et al., 1998). They would have covered about one-half of the total land area of the United States and included diverse hydrological systems that differ widely in the natural and human factors that affect water quality.

During Cycle I (as indicated in Figure 1-1 and Table 1-1), the study units were divided into three groups that were studied on a rotational schedule. The approximately decadal study cycle included three to five years of intensive data collection and analysis, preceded by an initial planning period and followed by several years of report preparation and low-level assessment activities (Gilliom et al., 1998). Thus, about one-third of the Cycle I study units were in the intensive data collection phase at any given time (Miller and Wilber, 1999).

To maximize their utility, study unit investigations are designed to meet the national synthesis requirements for consistent and integrated information, yet to remain flexible enough to meet individual, local needs of study units for assessing water quality. Each study unit investigation in Cycle I consisted of four interrelated components (Gilliom et al., 1995; Miller and Wilber, 1999):

INTRODUCTION 27

1. *Retrospective analysis.* This review and analysis of existing water quality data provided a historical perspective on water quality throughout the study unit. It formed the basis for addressing what was already known and aided the design of further investigations with respect to current water quality conditions, trends and changes, and understanding causes and effects (e.g., USGS, 1997).

2. *Occurrence and distribution assessment.* This component built on the findings of the retrospective analysis and characterized the broad-scale geographic and seasonal distributions of current water quality conditions in a study unit in relation to major point and nonpoint contaminant sources and natural or background conditions. This assessment was also used to help fill gaps in existing data for each study unit. Assessment features, including physical, chemical, and biological measurements, media sampled, and spatial and temporal resolution, were designed to be consistent among the study units. Results of this assessment were also used to identify the most important questions in the study unit area about sources, transport, fate, and effects to be addressed by the case studies (see below).

3. *Trend and change assessment.* This assessment focused on long-term (decadal-scale) trends and changes in water quality conditions. This component led to new questions about causes and effects in each study unit and identified changes that needed to be made in the periodic intensive study phases. Future trend and change assessments are intended to be based on data collected between intensive study phases and on data from subsequent three-year intensive phases.

4. *Case (synoptic) studies of sources, transport, fate, and effects.* Case studies were used to improve understanding of selected questions about sources, transport, fate, and effects that arise from all aspects of NAWQA investigations and often lead to changes in assessment approaches over time.

As noted previously, the occurrence and distribution assessment was the primary focus during Cycle I for all study units because it provides the foundation of data and information needed for later study unit investigation components.

## Surface Water

Because the national study design for surface water and groundwater studies in Cycle I has been described in detail (Gilliom et al., 1995, 1998), only a brief summary is provided here. The surface water design focuses on water quality conditions in rivers and streams (collectively referred to as streams) of each study unit. Lakes, reservoirs, and coastal waters were not addressed as part of the national design in Cycle I, although they are selectively investigated in some study units (e.g., Callender and Van Metre, 1997). Furthermore, while the surface water sites located throughout NAWQA included a wide range of stream sizes, types, and land-use settings in major regions of the nation, they were not selected to be a statistically representative sample of streams in the United States

(Miller and Wilber, 1999). Rather, the NAWQA study design for surface water investigations is somewhat biased toward sampling a greater prevalence of small (17 to 1,243 km$^2$) and large basins (1,244 to 221,497 km$^2$) with large proportions of agricultural and urban land compared to all similarly sized basins in the United States. The surface water study design used the following three interrelated components:

1. Water-column studies focused on assessing physical and chemical characteristics, including major ions, nutrients, organic carbon, dissolved pesticides, and suspended sediment, and related these characteristics to hydrologic conditions, sources, and transport.

2. Bed sediment and tissue studies were the primary means by which trace elements and hydrophobic organic contaminants (which tend to associate with particles and accumulate in biological tissues rather than be dissolved in water) were initially assessed in Cycle I of NAWQA.

3. Ecological studies evaluated the impacts of land use and the effects of physical and chemical characteristics of water on aquatic biota in streams, such as benthic macroinvertebrates, algae, and fish communities.

All three study components relied on coordinated sampling of varying intensity and scope at two general types of sites: integrator and indicator (Gilliom et al., 1995). In brief, integrator sites were chosen to represent water quality conditions of streams in heterogeneous large basins that are commonly affected by complex combinations of land-use settings, point and nonpoint sources of contaminants, and natural influences. Data from integrator sites provide a general check on the persistence of water quality influences evident at indicator sites associated with specific environmental settings (see more below) and are used for water budget and contaminant transport assessments. In contrast, indicator sites were chosen to represent water quality of streams in relatively homogeneous and usually smaller basins associated with specific environmental settings (e.g., a specific combination of geologic setting and land use). Thus, the water quality at indicator sites is influenced primarily by the targeted environmental setting. Furthermore, there are two special types of indicator sites known as reference and point source sites. Reference indicator sites were located downstream of undisturbed drainages that were chosen to represent background conditions within a basin. Point source indicator sites were located downstream of specific major point sources of contamination that may result in regionally significant water quality effects. They were usually paired with a site upstream of the point source.

The sampling strategy for water column, bed sediment and tissue, and ecological studies in Cycle I was based on monitoring at a combination of integrator and indicator sites (Gilliom et al., 1995). A few selected sites underwent intensive sampling for all water quality characteristics of concern related to all three components. Additional sites were included progressively and underwent more specialized but less frequent sampling. The most complete data set for the three components of a surface water investigation were collected at a selected core of three to five integrator sites and four to eight indicator sites in each study unit, which together constitute the fixed-site monitoring network for regular collection of samples over time (Miller and Wilber, 1999). In each study unit, a subset of two to five sites, usually one integrator site and two to four indicator sites, was sampled more intensively than the rest, and these were the only sites for which water samples were routinely analyzed for pesticides. Samples were also collected for low-level analyses of volatile organic compounds (VOCs).

## Groundwater

The national study design in Cycle I for groundwater focused on water quality conditions in major aquifers and in recently recharged shallow groundwater associated with present and recent human activities (Miller and Wilber, 1999). However, the agricultural and urban settings focus of the Cycle I NAWQA groundwater design was similar to the surface water investigations. Whereas stream water quality is highly variable over time, groundwater quality is determined primarily by chemical characteristics that tend to vary more spatially than temporally. Thus, the general groundwater monitoring approach in Cycle I focused on spatial characterization. Specifically, groundwater quality was assessed using three primary study components:

1. Aquifer or "study unit" surveys assessed the quality of water in the major aquifer systems of each study unit by primarily sampling existing wells.

2. Land-use studies used observation wells to assess the quality of recently recharged shallow groundwater (generally less than 10 years old) associated with regionally extensive combinations of land uses and hydrogeological conditions.

3. Flowpath studies used transects and groups of clustered, multilevel observation wells to examine specific relations among land-use practices, groundwater flow, contaminant occurrence and transport, and surface water and groundwater interactions.

Generally, each Cycle I aquifer survey and land-use study consisted of sampling about 30 randomly selected sites (wells or springs) in the geographic area and aquifer zone targeted for the specific study, with one sample collected from most of the sites (Gilliom et al., 1998). A total of 114 aquifer surveys and 105 land-use studies were completed in the 51 study units during Cycle I (Robert Gilliom, USGS, personal communication, 2001). Of the 105 land-use studies, 67 targeted agricultural settings, 34 targeted urban settings, and 4 targeted forested settings. One or two flowpath studies were planned in each study unit during the first cycle and were viewed as the potential beginning of long-term studies that would lead to more intensive groundwater investigations (Gilliom et al., 1995).

However, the flowpath study component of many study units in Cycle I was dropped because of a lack of background information and inadequate financial resources (Mallard et al., 1999).

## Selection of Target Analytes and Contaminants[1]

During Cycle I of NAWQA, the USGS selected a wide range of physical, chemical, and biological parameters to monitor in a nationally consistent manner (Miller and Wilber, 1999). In general, the selection of target analytes and contaminants was based on their relevance to important water quality issues and on the existence of appropriate analytical methods. Water quality measurements included field measurements of streamflow, pH, temperature, dissolved oxygen, and specific conductance and laboratory analyses of major ions, nutrients, trace elements, organic carbon, pesticides, and VOCs (Gilliom et al., 1995, 1998). Miller and Wilber (1999) noted that for some water quality issues (e.g., nutrient enrichment, acidification, salinity, sedimentation) the choice of target analytes is a relatively simple task and their analysis is relatively inexpensive. In contrast, selection of target analytes for chemical contaminants such as pesticides and VOCs is much more difficult because of the large number of substances to consider and their relatively high unit cost of measurement. Descriptions of biological communities were considered essential for an overall assessment of water quality resources and were made by assessing three taxonomic groups (fish, invertebrates, and algae) and habitat conditions. Chapter 3 of this report includes a discussion of the role of Cycle II of NAWQA in identifying and assessing emerging contaminants.

## National Synthesis

The national assessment component of NAWQA accomplishes two primary goals (Miller and Wilber, 1999). First, the regular accumulation of consistent and comparable water quality assessments for some of the largest, diverse, and important hydrological systems of the nation stands by itself as a major contribution to the knowledge of regional and national water quality. Second, the NAWQA national synthesis builds on and expands the findings from individual study units by combining them with other historical information (e.g., land use) reported by the USGS and other agencies and researchers. National synthesis teams produce regional and national assessments for priority water quality topics by comparative analysis of these study unit findings. National synthesis efforts focused on pesticides and nutrients began in 1991 and continue to this day, with

---

[1]Also collectively referred to as target "constituents" in some NAWQA reports. However, the committee generally uses the term "contaminant" throughout this report.

INTRODUCTION                                                                                                           31

several related USGS reports and other documents already having been published (e.g., Kolpin et al., 1998; USGS, 1999) and made available on the Internet. Similarly, national synthesis of data on VOCs began in 1994, trace elements in 1997, and ecological synthesis in 1999 (see Chapter 3 and 8 for further information on the national synthesis topics and teams).

## INITIAL ASSESSMENT OF NAWQA RESULTS AND ACCOMPLISHMENTS

As part of its statement of task and to help put the remainder of this report into proper context, the committee was charged to conduct an initial assessment of the general accomplishments of the NAWQA program to date (i.e., essentially Cycle I). More specifically, to perform such an assessment, the committee was tasked to engage in discussions—during official committee functions and independently—with USGS-NAWQA program scientists and other persons and organizations that routinely interact with NAWQA personnel and use or contribute to their data (i.e., "NAWQA users"). In this regard, at four of its five scheduled meetings the committee heard presentations and had discussions with several USGS-NAWQA personnel, including the chief hydrologist of the USGS, the NAWQA chief, NAWQA national synthesis chief, three national synthesis topic chiefs (VOCs, pesticides, and ecological synthesis), two study unit chiefs (Santa Ana Basin and Long Island-New Jersey Coastal Drainages), and other NAWQA scientists. In addition, the committee also heard presentations and had discussions with several NAWQA users and collaborators at most of its meetings. Outside of the scheduled meetings, however, all committee members met with or had phone discussions with various USGS-NAWQA staff and many NAWQA users in support of their deliberations and work on this report.

As required, the committee also carefully reviewed several iterations of draft internal reports by USGS-NAWQA that outline opportunities and strategies to improve NAWQA as it enters its second decade as a national monitoring program. However, the committee's overall assessment of the NAWQA program went beyond these stipulated activities and includes an abbreviated (by necessity) overview of the breadth, availability, and usefulness of NAWQA publications and Internet-based information to date. These are summarized below and discussed more extensively in Chapter 6 of this report. Several representative accomplishments and results of NAWQA to date are also provided to illustrate the widespread importance and success of this national monitoring program.

### Summary of Data and Information Dissemination in NAWQA

As of February 2001, almost 1,000 NAWQA-related documents have been published or otherwise made publicly available. These can be divided into several categories of documents including circulars, fact sheets, Open-File Reports,

Water-Resources Investigation Reports, conference proceeding papers, journal articles, and books or chapters. In addition, newsletters, academic theses, pamphlets, and other miscellaneous materials have been generated. These documents can be further classified as to whether they are generated at the study unit level (e.g., 36 individual study unit circulars from Cycle I have been published); are based on findings from national synthesis topics (e.g., *The Quality of Our Nation's Waters: Nutrients and Pesticides;* USGS, 1999); are technical documentation of study design, field protocols, and methods comparisons (Koterba et al., 1995); or are of national-level interest and outreach (e.g., *Review of Phosphorus Control Measures in the United States and Their Effects on Water Quality*, Litke, 1999). Much of this published information is also available through the Internet, and many of the data collected from the first 36 Cycle I study units are contained in the NAWQA Data Warehouse (an Oracle database) that is also available on the Internet.

As noted above, Chapter 6 provides further information and discussion of such NAWQA products as related to the communication of NAWQA data and findings to various users.

## Representative Accomplishments of Cycle I of NAWQA

The NAWQA program has been remarkably productive in its first decade of national monitoring. There are numerous accomplishments, important to every study unit, but also many that are recognized accomplishments at the national scale. As a sampling, the committee presents below 10 examples of the program's representative accomplishments that illustrate the breadth of contributions NAWQA has already made, and continues to make, to the nation.

### *Identification of Unexpectedly Frequent Occurrences of Pesticides in Urban Streams*

In Cycle I of NAWQA, water samples from streams flowing through agricultural and urban landscapes were analyzed for herbicides and pesticides. One result from these studies, which was surprising to many, was the prevalence of insecticides in urban streams (USGS, 1999). In contrast to agricultural landscapes, urban areas had not traditionally been considered important contributors to pesticide contamination of waters. NAWQA researchers documented widespread occurrence of insecticides that are commonly used in homes, gardens, and commercial areas. The most commonly occurring insecticides were diazinon, carbaryl, chlorpyrifos, and malathion. These occurred with higher frequency and oftentimes at higher concentrations in urban than in agricultural areas. Most urban streams sampled had insecticide concentrations exceeding water quality guidelines for health of aquatic organisms in 10 to 40 percent of samples collected throughout the year. Some herbicides commonly used on lawns, road margins, and golf courses were also detected frequently. Pesticides in urban streams most commonly occurred in mixtures, similar to agricultural areas, rather than as isolated occurrences of a single pesticide.

## Integrating Biological Assessments into Water Quality Monitoring

Previous USGS water quality monitoring has been directed almost exclusively to measures of physical and chemical aspects of water quality. During Cycle I of NAWQA, USGS scientists began a program of biological assessment that was integrated with the more traditional physical and chemical measures. As noted later in this report (Chapter 5), this data set will provide a valuable contribution to the ongoing debate on how best to do biological assessment and what such assessment tells us about water quality that was not evident from other methods. In addition, some interesting patterns were observed during the early Cycle I analyses. For example, in the Yakima River Basin (Cuffney et al., 1997), condition of algal and invertebrate communities declined precipitously above a certain value of an agricultural intensity index and then showed little further decline. This suggests a threshold response to increasing intensity of agricultural activity, which has implications for the design of both monitoring and restoration programs. Future monitoring efforts can be directed toward better documenting that part of the response curve nearest the threshold (i.e., assessing conditions when biological communities are affected most rapidly). The USGS has used this approach in its Cycle II design for assessing impacts of urban expansion (see Chapter 5). The existence of a threshold also provides a potential mechanism for prioritizing sites for restoration. Dollars invested in creating improvements in conditions at sites that are near the threshold are likely to provide a greater payoff in improved biological conditions than dollars invested in sites that are well beyond the threshold. At the most degraded sites, a much greater investment may be needed to improve conditions such that there is measurable improvement in biological integrity.

## Ecological and Human Health Issues

Potential impact on the reproductive, nervous, and immune systems from low-level exposure to environmental contaminants has become a growing concern for the protection of wildlife, aquatic organisms, and human health. The assessment of substances that may exert potential "endocrine disruption" health effects is a focus of the Food Quality Protection Act of 1996. NAWQA data have provided the first significant glimpse of potential associations among groups of pesticides and hormones in fish on a national scope. These early studies show a strong association of the highest pesticide concentrations in surface waters—though typically below suspected aquatic toxicity levels—with the lowest values of a key hormone ratio, which suggests impacts on endocrine functioning

(Goodbred et al., 1996). This is an important early signal that warrants further research.

*Emerging and Unexpected Chemical Contaminants in Water: MTBE*

Developing a list of chemical contaminants of concern that affect the nation's water quality is an ever-changing target. Each year, industry produces thousands of new chemicals that may be released intentionally or inadvertently into the environment. Furthermore, established chemicals occasionally may create unforeseen environmental problems or be of greater concern for ecological or human health than originally thought. These emerging contaminant concerns go far beyond the monitoring of currently regulated chemicals. NAWQA, through its broad-spectrum approach to contaminants in water (and to a lesser extent in biota and sediments), is on the front line to assess these new and emerging chemicals. Many of the analytical methods used in NAWQA are multiresidue methods (i.e., they can detect a broad spectrum of contaminants in a single sample). While NAWQA uses a list of target chemical compounds for which every sample is analyzed, USGS scientists often evaluate their samples and analysis results for nontarget compounds that may be present or look for new or unexpected chemical occurrence (e.g., through the use of computerized searches of mass spectral libraries).

In the 1970s, methyl *tert*-butyl ether (MTBE) was not generally regarded as a water contaminant, but as a compound that could help to produce cleaner air by being added to gasoline. In the 1990s, however, MTBE was discovered to be widely occurring in NAWQA and USGS Toxic Substances Hydrology Program samples (e.g., Delzer et al., 1996; Squillace et al., 1996; Zogorski et al., 1997) and was subsequently added to the USGS list of VOCs to be monitored. These data have played an important role in the U.S. Environmental Protection Agency (EPA) and several states in making a more thorough review of MTBE's potential impact on drinking water. In fact, these data showed that as an unintended consequence of MTBE use, shallow groundwater, particularly in urban areas, was showing widespread albeit low-level contamination. EPA and USGS have since entered into a jointly funded program to conduct more detailed monitoring for MTBE in some critical states where it is suspected to be a greater problem. For example, they have found that about nine percent of community drinking water systems in New England had detections of MTBE. These findings have prompted new state policies and monitoring as well.

*Assessment of Linkages Between Changes in Phosphorus Loads and Clean Water Act Policy*

One of the major accomplishments of NAWQA to date is the use of historical data on phosphorus loadings to evaluate the effects of various phosphorus control measures implemented in response to the Clean Water Act (Litke, 1999).

# INTRODUCTION

Phosphorus is a plant nutrient and enriched phosphorus levels in surface water can accelerate the growth of algae and other plants, resulting in eutrophic conditions in lakes and streams. Such conditions can harm the aquatic environment and impair recreational fishing and other uses. Phosphorus inputs to the environment began to increase since the 1950s as the use of fertilizers and phosphate-based laundry detergent increased. Eutrophication was recognized as a primary water quality concern during the 1960s, and reducing eutrophic conditions in surface water was one of the goals of the Clean Water Act. Some of the actions taken to reduce phosphorus levels included upgrading wastewater treatment plants, state-level bans on phosphate detergents, and reducing runoff from cropland.

As part of the (NAWQA) National Synthesis for nutrients, Litke (1999) reviewed the history of phosphorus controls in the United States and demonstrated how retrospective data collected by the NAWQA program might be used to evaluate the effect of phosphorus controls on water quality. More specifically, data from NAWQA are representative of a variety of phosphorus control measures and therefore may be used to evaluate the effects of various control strategies. To date, data show that phosphorus concentrations are reported to be decreasing in many NAWQA study units, largely a result of the reduction in the use of phosphate detergents and phosphorus limits at treatment plants. However, more analysis needs to be done to explain why phosphorus concentrations have not declined in some basins.

## Development of SPARROW

SPARROW (Spatially Referenced Regressions on Watershed Attributes) is a watershed scale model with a mechanistic basis and is a major contribution to water quality modeling tools and literature. SPARROW was developed by the USGS (Smith et al., 1997), although not under the NAWQA program. However, NAWQA's highly visible use of SPARROW is helping to make the model increasingly known and is enabling many parties to use it. The watershed model has many desirable characteristics including consideration of multiple sources of nutrients, a good balance between land-to-water processes and in-stream processes, its ability to be validated with monitored data, and the fact that it quantifies uncertainties. SPARROW has been used to assess sources of nutrients in the nation's watersheds (Smith et al., 1997), to predict total nitrogen transport in rivers of the conterminous United States (Smith and Alexander, 2000), to predict the delivery of nitrogen to the Gulf of Mexico that contributes to hypoxia (Alexander et al., 2000), and to predict total nitrogen in the monitoring sites in the Chesapeake Bay watershed. SPARROW has the potential to become one of the most commonly applied models in support of the EPA's Total Maximum Daily Load (TMDL) Program because it employs watershed parameters that may be estimated using generally available data.

## Hyporheic Zone Work Throughout NAWQA

The hyporheic zone, or surface water-groundwater ecotone, is the interface between groundwater and surface water (Gibert et al., 1990; Vervier et al., 1992). In this region, active and dynamic exchange of water, organisms, and chemicals occurs between the surface water and the adjacent groundwater system (Gibert et al., 1990; Triska et al., 1989; Vervier et al., 1992). Thus, the hyporheic zone is an important component for understanding surface water quality and near-surface groundwater quality, and NAWQA researchers are advancing our knowledge of hyporheic processes and their effects on water quality. For example, Hinkle et al. (2001) showed that calculation of groundwater solute loads to large streams based on regional groundwater geochemistry, without accounting for hyporheic zone biogeochemical transformations, could result in significant errors. Another excellent example of surface water-groundwater-hyporheic zone water quality work emerging from Cycle I of NAWQA is that of McMahon and Bohlke (1996). They showed that along the South Platte River, denitrification and surface water-groundwater mixing within the alluvial aquifer sediments may greatly decrease the nitrate added to streams by discharging groundwater.

## Information Dissemination in NAWQA

Although most of the NAWQA budget and effort is devoted to data collection and interpretation, it is the reporting of the program's findings that is most critical for its widespread use and ultimate success. In this regard, the NAWQA program is generating a tremendous amount of data and information that are of interest to researchers, resource management and regulatory agencies, and the general public. A wide range of information has been made available, including program history and design, sampling and analysis protocols, study unit findings, and interpretations at both the study unit and the national scales. NAWQA information is communicated to a wide audience through reports, databases, the Internet, and other digital products. Information generated by NAWQA is already being used by a number of states to assist them in designing and implementing additional water quality programs (see Chapters 6 and 7 for further discussion).

## Improved Methods and Detection Levels for Chemical Contaminants

NAWQA program personnel have worked to refine and improve existing analytical methods while simultaneously lowering their detection and reporting levels (i.e., concentrations). These efforts have also led to improvements outside the laboratory to include field sampling and processing and extraction of samples. However, many of these improvements require well-trained field personnel (such as USGS staff) to collect samples. The improved, lower-level reporting concentrations are another important contribution. For example, as one tries to gain an increased understanding of chemical contaminant occurrence, fate and transport, and cause-and-effect relationships, it is important to know what the real concentration of a contaminant is in the environment—that is, to distinguish between low-level and "zero" occurrence (or censored "less than" data) of chemical contaminants. Other organizations have adopted USGS protocols to achieve lower detection levels for new studies.

*Improved Coordination and Collaboration with Other Federal and State Agencies*

NAWQA has become a model of an effective, collaborative federal program—an attribute that policy makers always stress, but seldom achieve. Within the USGS, NAWQA has successfully integrated its program with the Toxic Substances Hydrology Program, the National Research Program, and the District-State Cooperative Programs, for example. NAWQA has established some exemplary relations with EPA and state governments. It has placed USGS staff within key EPA offices. The EPA and both state and local agencies have found NAWQA of such value that they have furthered the symbiosis by providing funding to USGS for NAWQA to help meet their additional information needs (see Chapter 7 for further discussion).

These examples are but a small cross section of NAWQA accomplishments. The committee's initial assessment finds that NAWQA is a well-organized, managed, and implemented program. It has significantly contributed to understanding the quality of the nation's water resources and the natural and anthropogenic factors that affect their quality. In this report, the committee provides guidance and recommendations for program improvements in Cycle II. In a program of this scope, there is always room for improvements and new efficiencies. However, the committee is duly impressed with the significance and magnitude of NAWQA's accomplishments in Cycle I. NAWQA clearly has provided the nation with a significant body of new knowledge to better understand and manage our vital water resources. The program has become an exemplary institution, illustrating that sound science can be applied at a national scale to resource assessment. Finally, NAWQA has assumed a vital leadership role, helping to improve environmental monitoring in many agencies from the federal to the local levels, both by its example and by technical assistance to others.

## INTRODUCTION TO CYCLE II OF NAWQA

As NAWQA neared the end of its first decade of nationwide study, the USGS internally organized the NAWQA Planning Team (NPT) in January 1997 to evaluate the overall success of the program and make recommendations for NAWQA design as it begins its second cycle (see Mallard et al., 1999). Overall, the NPT lauded NAWQA's valuable contributions to the understanding of the quality of the nation's waters and concluded that the principles established during the first decade of the program should guide NAWQA during its second decade. The NPT provided 10 major conclusions and recommendations for the design of Cycle II of NAWQA, the most important being a recommended shift in the overall emphasis of the program from the occurrence and distribution of selected contaminants to enhanced efforts toward understanding and explaining the processes controlling water quality. This shift in program emphasis is reflected in the committee's statement of task and is the primary subject of Chapter 5 of this report. The other major theme pervading the NPT report is the realization that program-wide design changes are needed for the second cycle of NAWQA to maintain and enhance the program in the face of funding shortfalls that are not likely to improve (and may worsen) in the immediate future. One major consequence of these budgetary constraints was a strong recommendation for the reduction in the total number of study units investigated in Cycle II rather than reducing the extent of monitoring and assessment activities.

In June 1999, the USGS internally formed the NAWQA Cycle II Implementation Team to develop a strategy for implementing the second cycle of NAWQA studies. Furthermore, Cycle II investigations must be based on the recommendations of the NPT, budget and management considerations, and experiences learned from implementing Cycle I of NAWQA. The first of two reports by the NIT describes its recommendations for the prioritization and selection (reduction) of study units for the second cycle of NAWQA (Gilliom et al., 2000a). Based largely on this report, the USGS recently decided to reduce the total number of study units in Cycle II to 42, not including the HPGW, which is being monitored intensively for six years and is the financial equivalent of two regular study units. An analysis of how this reduction was accomplished and its implications can be found in Chapter 2 of this report.

The second and broader report of the NIT describes the design and implementation strategy for Cycle II investigations in the (now reduced in number) NAWQA study units. As noted previously, several iterations of this internal report (Gilliom et al., 2000b; see also Appendix A) were provided to the committee to assist its deliberations and ultimately helped determine (in conjunction with the statement of task) the report's overall structure and organization.

## Coordination Efforts

At all levels of the program, NAWQA benefits from information exchange and coordination activities that occur through several mechanisms (Miller and Wilber, 1999). First, each study unit is associated with and assisted by a liaison committee to help ensure that the water quality information produced by the program is relevant to regional and local interests. These liaison committees are comprised of non-USGS personnel that bring a balance of technical and management interests. Typically represented organizations include federal, state, interstate, and local agencies, Indian nations, and universities. At the national level, the NAWQA Advisory Committee was formed in 1991 to improve interaction between the USGS and parties with an interest in the success of the NAWQA program. The national liaison structure was changed in 1999. In 2000, the NAWQA National Liaison Committee was established to replace the dissolved NAWQA Advisory Council. Chapter 7 provides a more detailed discussion of the history and role of NAWQA's national liaison efforts.

The USGS and NAWQA have also established liaison positions with the EPA's Office of Ground Water and Drinking Water and Office of Pesticide Programs to provide a stronger and more timely linkage with the information needs of these two groups. Chapter 7 also describes a variety of federal, state, and local agencies and programs with which NAWQA collaborates. Finally, as described earlier in this chapter, the NRC's Water Science and Technology Board has provided advice to the USGS four times in the past—from NAWQA's inception to the present.

## CONCLUSIONS

An initial assessment of Cycle I of NAWQA and its representative accomplishments to date finds that it has evolved into a mature and respected national program, with hundreds of publications to its credit and many significant science and policy achievements related to the quality of our nation's ground- and surface waters for it to build upon. NAWQA has produced not only an unprecedented volume of quality data for use in the scientific community, but also unbiased information that is being used by decision makers, managers, and planners at all government levels. However, NAWQA program personnel have also worked to refine and improve existing analytical methods while simultaneously lowering their detection and reporting levels (i.e., concentrations). These efforts have also led to improvements outside the laboratory to include field sampling and processing and extraction of samples. The use of NAWQA information, and the linkages that many other organizations continually seek to make with NAWQA are an illustration of the important void that NAWQA has filled in the realm of the nation's water quality. NAWQA is also to be commended for striving for continual improvement and, to this end, has repeatedly asked for review and critical input from various stakeholders, interest groups, and the NRC. Despite their accomplishments and impressive legacy of quality reports, NAWQA staff will be increasingly challenged to plan and execute monitoring in Cycle II because of budgetary constraints. For these reasons, NAWQA must continue to review its efficiency and cost-effectiveness; NAWQA staff will have to apply the lessons learned from their first decade of national monitoring and find new ways to operate effectively. With such continued diligence and improvement, NAWQA should be able to meet the challenges and goals that Congress and the nation have asked of it.

# REFERENCES

Alexander, R. B., R. A. Smith, and G. E. Schwarz. 2000. Effects of stream channel size on the delivery of nitrogen to the Gulf of Mexico. Nature 403:758-761.

Callender, E., and P. C. Van Metre. 1997. Reservoir sediment cores show U.S. lead declines. Environmental Science and Technology 31:424a-428a.

Cuffney, T. F., M. R. Meador, S. D. Porter, and M. E. Gurtz. 1997. Distribution of Fish, Benthic Invetebrate, and Algal Communities in Relation to Physical and Chemical Conditions, Yakima River Basin, Washington, 1990. U.S. Geological Survey Water-Resources Investigations Report 96-4280. Reston, Va.: U.S. Geological Survey.

Delzer, G. C., J. S. Zogorski, T. J. Lopes, and R. L. Bosshart. 1996. Occurrence of the Gasoline Oxygenate MTBE and BTEX Compounds in Urban Stormwater in the United States, 1991-1995. U.S. Geological Survey Water-Resources Investigations Report 96-4145. Reston, Va.: U.S. Geological Survey.

Gibert, J., M. J. Dole-Olivier, P. Marmonier, and P. Vervier, P. 1990. Surface water/groundwater ecotones. Pp. 199-225 in Naiman, R. J., and H. Decamps (eds.) Ecology and Management of Aquatic-Terrestrial Ecotones. Man and the Biosphere Series. Volume 4. Paris: United Nations Educational, Scientific, and Cultural Organization.

Gilliom, R. J., W. M. Alley, and M. E. Gurtz. 1995. Design of the National Water Quality Assessment Program—Occurrence and Distribution of Water-Quality Conditions. U.S. Geological Survey Circular 1112. Sacramento, Calif.: U.S. Geological Survey.

Gilliom, R. J., D. K. Mueller, and L. H. Nowell. 1998. Methods for Comparing Water-Quality Conditions Among National Water Quality Assessment Study Units, 1992-1995. U.S. Geological Survey Open-File Report 97-589. Sacramento, Calif.: U.S. Geological Survey.

Gilliom, R. J., K. Bencala, W. Bryant, C. Couch, N. Dubrovsky, D. Helsel, I. James, W. Lapham, M. Sylvester, J. Stoner, W. Wilber, D. Wolock, and J. Zogorski. 2000a. Prioritization and Selection of Study Units for Cycle II of NAWQA. U.S. Geological Survey NAWQA Cycle II Implementation Team. Draft for internal review. Sacramento, Calif.: U.S. Geological Survey.

Gilliom, R. J., K. Bencala, W. Bryant, C. A. Couch, N. M. Dubrovsky, L. Franke, D. Helsel, I. James, W. W. Lapham, D. Mueller, J. Stoner, M. A. Sylvester, W. G. Wilber, D. M. Wolock, and J. Zogorski. 2000b. Study-Unit Design Guidelines for Cycle II of the National Water Quality Assessment (NAWQA). U.S. Geological Survey NAWQA Cycle II Implementation Team. Draft for internal review (11/22/2000). Sacramento, Calif.: U.S. Geological Survey.

Gilliom, R. J., P. A. Hamilton, and T. L. Miller. 2001. The National Water-Quality Assessment Program—Entering a New Decade of Investigations. U.S. Geological Survey Fact Sheet 071-01. Reston, Va.: U.S. Geological Survey. Available online at http://water.usgs.gov/pubs/FS/fs-071-041/pdf/fs07101.pdf.

Goodbred, S. L., R. J. Gilliom, T. S. Gross, N. P. Denslow, W. B. Bryant, and T. R. Schoeb. 1996. Reconnaissance of 17β-Estradiol, 11-Ketotestosterone, Vitellogenin, and Gonad Histopathology in Common Carp of United States Streams: Potential for Contaminant-Induced Endocrine Disruption. U.S. Geological Survey Open-File Report 96-627. Sacramento, Calif.: U.S. Geological Survey.

Hinkle, S., J. Duff, F. Triska, A. Laenen, E. Gates, K. Bencala, D. Wentz, and S. Silva. 2001. Linking hyporheic flow and nitrogen cycling near a large river in the Willamette Basin, Oregon. Journal of Hydrology 244(3-4):157-180.

Hirsch, R. M., W. M. Alley, and W. G. Wilber. 1988. Concepts for a National Water-Quality Assessment Program. U.S. Geological Survey Circular 1021. Denver, Colo.: U.S. Geological Survey.

Kolpin, D. W., J. E. Barbash, and R. J. Gilliom. 1998. Occurrence of pesticides in shallow groundwater of the United States: Initial results from the National Water-Quality Assessment Program. Environmental Science and Technology 32(5):558-566.

# INTRODUCTION

Koterba, M. T., F. D. Wilde, and W. W. Lapham. 1995. Ground-Water Data-Collection Protocols and Procedures for the National Water-Quality Assessment Program: Collection and Documentation of Water-Quality Samples and Related Data. U.S. Geological Survey Open-File Report 95-399. Reston, Va.: U.S. Geological Survey.

Litke, D. W. 1999. Review of Phosphorus Control Measures in the United States and Their Effects on Water Quality. U.S. Geological Survey Water-Resources Investigations Report 99-400. Denver, Colo.: U.S. Geological Survey.

Lynn, W. R. 1985. Letter to D. Peck, Director, USGS. Reston, Va.

Mallard, G. E., J. T. Armbruster, R. E. Broshears, E. J. Evenson, S. N. Luoma, P. J. Phillips, and K. R. Prince. 1999. Recommendations for Cycle II of the National Water Quality Assessment (NAWQA) Program. U.S. Geological Survey NAWQA Planning Team. U.S. Geological Survey Open File Report 99-470. Reston, Va.: U.S. Geological Survey.

McMahon, P. B., and J. K. Bohlke. 1996. Denitrification and mixing in a stream-aquifer system: Effects on nitrate loading to surface water. Journal of Hydrology 186:105-128.

Miller, T. L., and W. G. Wilber. 1999. Emerging drinking water contaminants: Overview and role of the National Water Quality Assessment Program. Pp. 33-42 in Identifying Future Drinking Water Contaminants. Washington, D.C.: National Academy Press.

NRC (National Research Council). 1986. National Water Quality Monitoring and Assessment. Report on a Colloquium Sponsored by the Water Science and Technology Board, May 21-22, 1986. Washington, D.C.: National Academy Press.

NRC. 1990. A Review of the USGS National Water Quality Assessment Pilot Program. Washington, D.C.: National Academy Press.

NRC. 1994. National Water Quality Assessment Program: The Challenge of National Synthesis. Washington, D.C.: National Academy Press.

NRC. 2001. Future Roles and Opportunities for the U.S. Geological Survey. Washington, D.C.: National Academy Press.

Smith, R. A., and R. B. Alexander. 2000. Sources of nutrients in the nation's watersheds. In Managing Nutrients and Pathogens from Animal Agriculture. Proceedings from the Natural Resource, Agriculture, and Engineering Service Conference for Nutrient Management Consultants, Extension Educators, and Producer Advisor. Camp Hill, Pa. March 28-30. Available online at http://water.usgs.gov/nawqa/sparrow/nut_sources/nut_sources.htm.

Smith, R. A., G. E. Schwarz, and R. B. Alexander. 1997. Regional interpretation of water quality monitoring data. Water Resources Research 33(12):2781-2798.

Squillace, P. J., J. S. Zogorski, W. G. Wilber, and C. V. Price. 1996. Preliminary assessment of the occurrence and possible sources of MTBE in groundwater in the United States, 1993-1994. Environmental Science and Technology 30(5):1721-1730.

Triska, F. J., V. C. Kennedy, R. J. Avanzino, G. Zellweger, and K. E. Bencala. 1989. Retention and transport of nutrients in a third-order stream in northwestern California: Hyporheic processes. Ecology 70:1893-1905.

USGS (U.S. Geological Survey). 1997. Pesticides in Surface Waters: Current Understanding of Distribution and Major Influences. U.S. Geological Survey Fact Sheet FS-039-97. Sacramento, Calif.: U.S. Geological Survey.

USGS. 1999. The Quality of Our Nation's Waters: Nutrients and Pesticides. U.S. Geological Survey Circular 1225. Reston, Va.: U.S. Geological Survey.

Vervier, P., J. Gibert, P. Marmonier, and M. J. Dole-Olivier. 1992. A perspective on the permeability of the surface freshwater-groundwater ecotone. Journal of the North American Benthological Society 11:93-102.

Zogorski, J. S., A. Morduchowitz, A. L. Baehr, B. J. Bauman, D. L. Conrad, R. T. Drew, N. E. Korte, W. W. Lapham, J. F. Pankow, and E. R. Washington. 1997. Fuel oxygenates and water quality. Chapter 2 in Interagency Assessment of Oxygenated Fuels. Washington, D.C.: Office of Science and Technology Policy, Executive Office of the President.

# 2

# Transition from Cycle I to Cycle II: Representativeness of Study Units

## INTRODUCTION

The National Water Quality Assessment (NAWQA) Program was designed to assess the status of and trends in the quality of the nation's groundwater and surface water resources and to link the status and trends with an understanding of the natural and human factors that affect the quality of water (Gilliom et al., 1995). Cycle I (1991-2001) of the program set out to accomplish this goal with plans to sample 60 study units (reduced to 59 in 1996)[1] that would cover more than half of the nation's land area and account for about 70 percent of the nation's drinking water use.[2] However, because of budgetary constraints, a total of eight study units that were slated for monitoring from 1997 to 2001 in Cycle I were never initiated. Thus, Cycle I of NAWQA included a total of 51 study units and the High Plains Regional Ground Water (HPGW) Study that was initiated in 1999 (see Chapter 1 for further information on Cycle I monitoring). Despite cutbacks, a major effort during Cycle I was the compilation and synthesis of study unit data to make inferences about "regional" and "national" water quality (e.g., USGS, 1999).

Continuing budget constraints have now dictated that the number of study units in Cycle II be reduced from 59 (for planning purposes) to 42, plus the HPGW study (see Figure 2-1 and Table 2-1). As in Cycle I, the Cycle II study units will

---

[1]The Northern New England Basins Study Unit merged with the Southeast New England Study Unit to form the New England Coastal Basins (NECB) Study Unit.

[2]It is important to note that water quality conditions are not necessarily studied throughout an entire study unit.

43

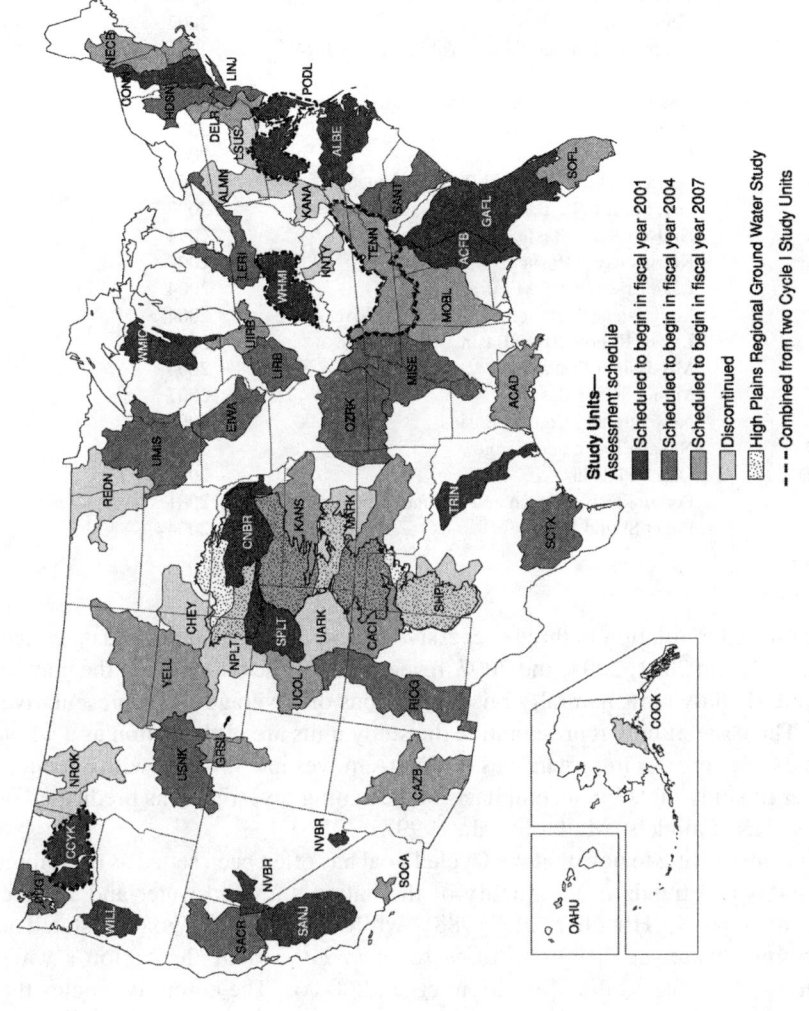

FIGURE 2-1 Planned Cycle II and discontinued Cycle I study units. SOURCE: Gilliom et al., 2001.

TABLE 2-1 Planned Cycle II NAWQA Study Units[a]

| Abbreviation | Name | Scheduled Starting Year |
|---|---|---|
| ACAD | Acadian-Pontchartrain Drainages | 2007 |
| ACFB | Apalachicola-Chattahoochee-Flint River Basins | 2001 |
| ALBE | Albemarle-Pamlico Drainages | 2001 |
| CACI | Canadian-Cimarron River Basins | 2007 |
| CAZB | Central Arizona Basins | 2007 |
| CCYK | Central Columbia Plateau-Yakima River Basin[b] | 2001 |
| CNBR | Central Nebraska Basins | 2001 |
| CONN | Connecticut, Housatonic, and Thames River Basins | 2001 |
| DELR | Delaware River Basin | 2007 |
| EIWA | Eastern Iowa Basins | 2004 |
| GAFL | Georgia-Florida Coastal Plain Drainages | 2001 |
| GRSL | Great Salt Lake Basins | 2007 |
| HDSN | Hudson River Basin | 2004 |
| KANS | Kansas River Basin | 2007 |
| LERI | Lake Erie-Lake St. Clair Drainages | 2004 |
| LINJ | Long Island-New Jersey Coastal Drainages | 2004 |
| LIRB | Lower Illinois River Basin | 2004 |
| MISE | Mississippi Embayment | 2004 |
| MOBL | Mobile River Basin | 2007 |
| NECB | New England Coastal Basins | 2007 |
| NVBR | Nevada Basin and Range | 2001 |
| OZRK | Ozark Plateaus | 2004 |
| PODL | Potomac River Basin and Delmarva Peninsula[c] | 2001 |
| PUGT | Puget Sound Basin | 2004 |

be monitored in rotation in three successive periods of high-intensity data collection that start in 2001, 2004, and 2007, respectively. The reduction in the number of Cycle II study units naturally raises questions of coverage and representativeness. The issue of how representative the study units are of the nation as a whole assumes even greater importance as NAWQA moves into Cycle II with a reduced number of study units but an emphasis on designing investigations predicated on process-based models (Mallard et al., 1999).

It is interesting to note that the Cycle I goal has often been stated as describing "the status and trends in the quality of the nation's ground-water and surface-water resources" (Hirsch et al., 1988), while the Cycle II goal is stated as describing "water-quality conditions for a *large part* of the nation's water resources" [emphasis added] (Gilliom et al., 2000a). The committee notes that the addition of the words "large part" is a significant and appropriate clarification of some earlier descriptions. In this regard, such inferences refer to the NAWQA study units, which are distributed across the nation but not clearly established to be "representative" of the nation. For the most part, recent written explanations of NAWQA results appear to be appropriately expressed; for example, Kolpin et

TABLE 2-1 Continued

| Abbreviation | Name | Scheduled Starting Year |
|---|---|---|
| RIOG | Rio Grande Valley | 2004 |
| SACR | Sacramento River Basin | 2004 |
| SANJ | San Joaquin-Tulare River Basins | 2001 |
| SANT | Santee River Basin and Coastal Drainages | 2004 |
| SCTX | South-Central Texas | 2004 |
| SOCA | Southern California Coastal Drainages[d] | 2007 |
| SOFL | Southern Florida Drainages | 2007 |
| SPLT | South Platte River Basin | 2001 |
| TENN | Tennessee River Basin[e] | 2007 |
| TRIN | Trinity River Basin | 2001 |
| UCOL | Upper Colorado River Basin | 2007 |
| UIRB | Upper Illinois River Basin | 2007 |
| UMIS | Upper Mississippi River Basin | 2004 |
| USNK | Upper Snake River Basin | 2004 |
| WHMI | White, Great and Little Miami River Basins[f] | 2001 |
| WILL | Willamette Basin | 2001 |
| WMIC | Western Lake Michigan Drainages | 2001 |
| YELL | Yellowstone River Basin | 2007 |

[a]Refer to Table 1-1 for a listing of states included in each planned Cycle I study unit (many of which are continued in Cycle II) and for definitions of those Cycle I study units that were discontinued or merged for Cycle II (see more below).
[b]Merged from CCPT and YAKI Cycle I Study Units.
[c]Merged from POTO and DLMV Cycle I Study Units.
[d]Called Santa Ana Basin (SANA) in Cycle I.
[e]Merged from UTEN and LTEN Cycle I Study Units.
[f]Merged from MIAM and WHIT Cycle I Study Units.

al. (1998) were careful to state inferences as 73 percent of the *sampling sites* or 54 percent of the wells *sampled*. Thus, NAWQA scientists should continue to endeavor to correctly state the geographic extent of scientific inferences.

This chapter assesses the coverage and representativeness of NAWQA study units and provides recommendations designed to help ensure that Cycle II remains as effective as Cycle I in assessing water quality status, trends, and understanding. More specifically, the chapter begins with an assessment of the representativeness of the nationwide study unit approach to NAWQA that is continued (albeit with a reduced number of study units) into Cycle II. Next, the importance of lakes, reservoirs, coastal waters, and permafrost and their coverage in Cycle II is evaluated. The adequacy of the Cycle II study design for monitoring groundwater and surface water is then assessed (see Chapters 3 to 5 for further information). The last section is a summary of the chapter's conclusions and recommendations.

This assessment of the coverage and representativeness of the reduced number of study units planned for Cycle II is fundamental to addressing the cross-cutting issues of extrapolation and aggregation of NAWQA data, posed in the committee's statement of task. If the reduced number of study units drastically altered the coverage and representativeness of NAWQA (e.g., by focusing only on urbanized areas), it would affect the potential for extrapolation and aggregation, as well as the potential for trend analysis and cause-and-effect studies.

## REPRESENTATIVENESS OF THE STUDY UNIT APPROACH

### Background

Representativeness depends upon the degree to which the study unit sites represent the variety of land uses and contaminant sources in the nation's watersheds and aquifers. In the language of the Cycle II design guidelines (Gilliom et. al., 2000b), "knowledge of land use and contaminant sources, natural characteristics of the land and hydrologic system, and our understanding of governing processes" have to be considered in judging representativeness. To develop a representative group of 42 study units from the original monitored or planned 59 study units in Cycle I, NAWQA program staff employed both quantitative (linear programming) and qualitative ("expert judgment") approaches. These approaches are discussed below.

### The Hydrologic Setting Regions Approach

The representativeness of the Cycle II study units with respect to hydrologic characteristics is addressed through the identification and delineation of hydrologic landscapes (Winter, 1995, 2000, 2001) and hydrologic setting regions throughout the United States (Wolock et al., 2000). Winter's concept is based on the idea that a single, simple physical feature—called a fundamental hydrologic landscape unit (FHLU)—can be described as the basic building block of all landscapes. This feature is an upland adjacent to a lowland separated by a steeper break in slope. The hydrologic system of a fundamental landscape unit consists of the movement of (1) surface water, controlled by the slopes and permeabilities of its surfaces; (2) groundwater, controlled by the hydraulic characteristics of its geologic framework; and (3) atmospheric water exchange, controlled by climate. This classification system is relevant to NAWQA because these hydrologic processes exert substantial control over water quality. The hydrologic setting region approach used in the NAWQA analysis combines Winter's three characteristics, plus temperature, in a geographic information system (GIS), and uses statistical analyses to define a total of 17 hydrologic regions (Wolock et al., 2000).

Details of the hydrologic landscape approach within NAWQA are presented in the NAWQA Cycle II Implementation Team (NIT) report by Gilliom et al.

(2000a). However, the three critical characteristics are (1) land surface form (relief, percentage of flat land in various portions of the region); (2) geologic texture (soil and bedrock permeabilities); and (3) climate (mean annual precipitation minus potential evapotranspiration for 1961 to 1990). Note that geologic texture does not directly consider lithology and mineralogy, both of which can profoundly influence water chemistry and, hence, water quality. It would therefore be quite possible to classify two aquifers as similar based on high permeabilities (e.g., a recent basalt and a cavernous limestone) although they might be quite dissimilar in terms of mineralogy or lithology and water quality. This is a shortcoming in the current classification methodologies.

These three characteristics were then combined with temperature (which can also profoundly affect water quality) and averaged for a total of 2,244 hydrologic cataloging units (HCUs). Wolock et al. (2000) then used GIS, principal components, and cluster analyses to operationalize Winter's hydrologic landscapes concepts to define 17 hydrologic setting regions for the 50 states. These regions were mapped over the entire nation and combined with the map of NAWQA Cycle I study units to help select sites for Cycle II that correspond to the occurrence of the hydrologic setting regions across the nation. Thus, hydrologic setting regions can be used to help achieve representativeness for key natural hydrologic factors as long as the initial set of NAWQA study sites provides adequate coverage of the 17 region types.

For a hydrologic setting to be significant in a particular study unit, it had to cover at least 10 percent of the study unit's total area. Furthermore, the selected Cycle II study units were chosen to cover 20 percent of the national extent of each hydrologic setting. (In addition, total drinking water use in the selected study units could not fall below 50 percent of the national drinking water use, as described further below.) Based on this exercise, 16 of the 17 hydrologic settings were included in the Cycle II study units. The only setting omitted (cold arid hills) occurs solely in Alaska but represents a very small percentage of the land area of the only Cycle I study unit located in Alaska (Cook Inlet Basin; see more below).

## Study Unit Reduction

The actual selection of 42 Cycle II study units from the 59 monitored or planned study units in Cycle I was accomplished through use of a linear programming approach (LPA) to ensure that the reduced number of study units would still represent at least 50 percent of the nation's drinking water by use, while reflecting a cross section of the nation's hydrologic settings (as discussed above) and ecological regions. Semiquantitative analyses (SQA) were also employed to ensure that the final group of 42 study units included the top 10 regions representing major contaminant sources (urban, agricultural, and natural), major aquifer systems, and sufficient ecological diversity (as reflected by the critical watershed

> **BOX 2-1**
> **Selection of Cycle II Study Units Using a Linear Programming Approach and Semiquantitative Analyses**
>
> The reduction from 59 (monitored or planned) study units to 42 for Cycle II was achieved using a linear programming approach (LPA) in conjunction with semiquantitative analyses (SQA) to account for factors not considered by LPA (Gilliom et al., 2000a). The LPA was used to optimize the selection of study units based on maximizing drinking water use and distributing the study units over a representative range of hydrologic setting regions. The former is closely correlated with population, while a representative range of hydrologic settings depicts regions with a broad range of natural transport and ecological characteristics. The LPA was also subject to three constraints: (1) for a hydrologic setting to be considered significant in a study unit, it had to comprise at least 10 percent of the study unit's area; (2) the selected study units had to include at least 20 percent of the national extent of each hydrologic setting; and (3) total drinking water use (groundwater public + groundwater domestic + surface water public) in the selected study units should not fall below 50 percent of national drinking water use. Only 16 of the 17 hydrologic settings were considered because one of the settings (cold arid hills) occurs only in Alaska, covering most of the state but just 2 percent of the COOK Study Unit. The initial LPA exercise produced 42 specific study units for continuation into Cycle II.
>
> SQA was subsequently employed to ensure that major contaminant sources (agricultural, urban, natural) and aquifer systems were represented. For each source category, the top 10 study units (of the original 59) were listed. To these three groups were added the study units required to represent coverage of the major aquifer systems. This exercise produced a "top-priority" group of 27 study units.

concept discussed below). Finally, those study units that had been selected for discontinuation were subsequently examined to ascertain if they possessed characteristics that would otherwise warrant their inclusion in the final 42. Box 2-1 describes the approach employed, and details can be found in Gilliom et al. (2000a).

A brief discussion of the critical watershed concept is warranted. This concept is derived from the publication *Rivers of Life* (The Nature Conservancy, 1998), which designated 327 small watersheds as "critical watersheds" to maxi-

The LPA was then rerun, constrained to included the 27 top-priority study units in the solution set; the consequent reduction in hydrologic setting coverage and drinking water use was then evaluated. The new solution set consisted of the 27 top-priority study units and 15 other units for a total of 42. The new set of 42 study units represented just a 6 percent reduction in the total national drinking water use and was deemed acceptable because it still represents 56 percent of the national drinking water use.

At this point, each of the original 59 study units was evaluated individually through SQA based on study unit coverage of ecologically critical watersheds. The NAWQA Cycle II Implementation Team also evaluated the 17 nonselected study units to determine if any had other characteristics that would merit their inclusion in the final 42 Cycle II study units. Five were so indicated: Albemarle-Pamlico Drainages (ALBE); Lower Tennessee River Basin (LTEN); Great and Little Miami River Basins (MIAM); White River Basin (WHIT); and Upper Colorado Basin (UCOL). To accommodate these, combining adjacent study units was considered. Ultimately, eight study units were combined into four: UTEN (Upper Tennessee River Basin) + LTEN; MIAM + WHIT; DLMV + POTO (Delmarva Peninsula + Potomac River Basin); and CCPT + YAKI (Central Columbia Plateau + Yakima River Basin). This produced a group of 29 study units. The LPA was run a third time using this group of 29 in the solution set and resulted in 13 additional study units to obtain a group of 42. Since the 13 nonselected study units contained the desirable UCOL and ALBE Study Units, these two were included in the final group of Cylcle II study units by dropping the Upper Arkansas River Basin (UARK) and the Allegheny and Monongahela Basins (ALMN) Study Units. These were discontinued because both have low agricultural and urban source contaminant rankings. When completed, the LPA and SQA approach produced a final group of 42 study units that account for about 61 percent of the national drinking water use.

mize coverage of "at risk" fish and mussel species. The watersheds were selected using the total number and percentage of threatened and endangered fish, mussel, and crayfish species. These watersheds represent 15 percent of the total number of eight-digit HCUs in the continental United States and include all known imperiled and vulnerable fish and mussel species. A value of 0 to 5 was assigned to each study unit based on the amount of coverage of critical watersheds (see Gilliom et al., 2000a, Appendix H). (Box 2-1 also describes how the critical watersheds were incorporated into the selection process.)

Through iterative use of LPA-SQA, the 42 planned Cycle II study units represent most of the regions that, regardless of the method of categorization, reflect the diversity in water quality and ecological characteristics of the nation's waters. The robustness of this approach can perhaps be best exemplified by examining the resulting suite of 42 study units selected for monitoring in Cycle II. Collectively, they account for (1) 16 of the nation's 17 hydrologic setting regions; (2) about 40 percent of the nation's land area (down 10 percent from the Cycle I plan); (3) 61 percent of the drinking water use (down about 10 percent from the Cycle I plan); (4) 80 percent of the nation's agricultural land; (5) about 80 percent of the nation's pesticide and fertilizer use; and (6) 76 percent (203 of 266) of the metropolitan statistical areas (MSAs[3]) within the continental United States have some overlap with at least one planned Cycle II study unit, while almost half (49 percent) of the total MSA area in the United States is within the planned Cycle II program area (Curtis Price, U.S. Geological Survey [USGS], personal communication, 2001).

However, a case could be made that none of the proposed 42 Cycle II study units includes a region of continuous or discontinuous permafrost, which would exert a strong control on the occurrence of groundwater. While the sole study unit covering this regime (Cook Inlet Basin; see Figure 2-1 and Table 1-1) has been slated for discontinuation, in terms of population and water use, this regime is not significant. However, its elimination does compromise the ability of the program to comment on the effects of climate change. The reason for this is that high-latitude regions, with their large expanses of ice, will respond quickly to warming or cooling and produce dramatic climatic (and hydrologic) changes (because of albedo changes), more so than temperate or tropical regions. This is referred to as "polar amplification" (Peter Fawcett, University of New Mexico, personal communication, 2001).

In addition, certain types of land use may not be represented adequately in the 42 Cycle II study units. For example, industries such as mining and forest products, whose operations can have a substantial influence on both surface water and groundwater quality, may not be well represented in either the original Cycle I or the reduced set of Cycle II study units, although several "mining indicator" lakes and reservoirs were sampled between 1992 and 2000 (Gilliom et al., 2000b). Similarly, the petrochemical industry, a potential major source of contamination, may not be well represented in Cycle II study units. Since major petrochemical areas were not specifically included as input information in determining the reduced set of study units, there is no way of easily determining how well or

---

[3]An MSA includes at least one city with 50,000 or more inhabitants or a U.S. Census Bureau-defined urbanized area (of at least 50,000 inhabitants) and a total metropolitan population of at least 100,000 (75,000 in New England; USCB, 2001). Twenty-three percent of the total land area of the United States is within an MSA (Curtis Price, USGS, personal communication, 2001).

poorly these areas are represented without further analysis. (In this regard, the committee notes that data from the Census Bureau's Census of Manufacturers or Annual Survey of Manufacturers could be used to help characterize the presence and intensity of various industrial sources of pollution in the planned Cycle II study units.)

Despite these potential shortcomings, the committee feels that the LPA-SQA methodology is an effective tool that provides a sound basis for reducing the number of study units in Cycle II without compromising the goals and objectives of the NAWQA program. Because of its ability to maximize results given certain constraints (i.e., diminishing resources) the LPA approach merits further application in other areas of the NAWQA program. For example, one area in which the NAWQA program might benefit from LPA is the area of sampling from drinking water intakes in reservoirs and lakes. Although the NAWQA program does not have an extensive limnological component, it is sampling at the drinking water intakes in about a dozen lakes and reservoirs (Timothy Miller, USGS, personal communication, 2000). If there are resources for sampling just a small or fixed number of lakes or reservoirs, LPA could be used to optimize the return on sampling—perhaps by finding the combination of lakes or reservoirs supplying the most people or having the largest drainage basins or the most agricultural lands, etc.

The LPA could also be used to assist selecting appropriate sites for surface water urban gradient studies (the "space-for-time" approach, discussed in Chapters 4 and 5). These planned Cycle II studies will be conducted in metropolitan areas with populations >100,000 persons. For these studies, the LPA could perhaps be used to help select sites that might represent different hydrologic settings and differing degrees of urban intensity to maximize results. Such expanded use of LPA within NAWQA might be an excellent way to optimize studies with already limited resources.

The committee notes that there are other useful methodologies that NAWQA scientists might consider when determining the representativeness of study units. For example, the urban intensity index is composed of variables (land cover, infrastructure, population, and socioeconomics) that reflect urbanization (McMahon and Cuffney, 2000). To assess the effects of urbanization, NAWQA watersheds with similar natural characteristics can be compared according to a gradient of the urban intensity index. One obvious advantage of such an approach is that a continuum of effect is quantified. A more detailed discussion of this index is in provided in Chapter 5.

## An Alternate Approach to Representativeness: EMAP

Since representativeness continues to be an important consideration in NAWQA as it enters its second decade of monitoring, it may be instructive to examine NAWQA in light of another monitoring program, the Environmental

Monitoring and Assessment Program (EMAP; see also Chapter 7) of the U.S. Environmental Protection Agency (EPA). This program, initiated at approximately the same time as NAWQA, took an altogether different approach to sampling despite the fact that the goals of both programs are similar. That is, EMAP also includes a national assessment of water quality. More specifically, EMAP was to "monitor the condition of the nation's ecological resources to evaluate the cumulative success of current policies and programs and to identify emerging problems before they become widespread or irreversible" (Messer et al., 1991). Specific objectives of EMAP were to (1) estimate current status, trends, and changes in selected indicators of the nation's ecological resources on a regional basis with known confidence; (2) estimate the geographic coverage and extent of the nation's ecological resources with known confidence; (3) seek associations between selected indicators of natural and anthropogenic stresses and indicators of the condition of ecological resources; and (4) provide annual statistical summaries and periodic assessments of the nation's ecological resources (EPA, 1991).

Ecological resources certainly encompass far more than just surface water and groundwater resources; however, the EMAP core research strategy document (EPA, 1997) indicates an overlap with NAWQA objectives in a number of topics concerned with water quality. The difference in the sampling approaches chosen for the two programs highlights the importance of how representative the NAWQA study units are. In particular, EMAP was based on a national probability sampling approach that was designed to permit statistical inference to a national or regional population. Probability sampling was not the basis for the selection of the NAWQA study units; rather, they were chosen to ensure that the most important national water quality issues in the nation's largest and most significant hydrologic systems were addressed (USGS, 1995). Failure to employ a probability sample to choose sampling sites does not prevent national inferences using NAWQA data, but it does underscore the need to express inferences correctly or to augment analyses with additional information or data to address the issue of representativeness. Although the absence of probability sampling does mean that NAWQA scientists cannot simply rely on statistical theory for inference or extrapolation to unsampled watersheds, such inferences are not prohibited if NAWQA scientists adequately consider and explain the representativeness of the sampling program to unsampled watersheds. Chapter 4 and particularly Chapter 5 discuss how inferences can be potentially extended to other areas.

It should also be noted that the use of a probability sample, as proposed in EMAP, does not directly support cause-and-effect studies or inferences, nor can trend analyses be accomplished any less cautiously. Furthermore, EMAP has never come to fruition as a national monitoring program (see Chapter 7); it remains in a regional and research mode. There are many unanswered questions about how to approach a probability sample for groundwaters as well as ecological resources.

## LAKES, RESERVOIRS, COASTAL WATERS, AND PERMAFROST

### Background

The NAWQA program, by design, does not sample many lakes (including the Great Lakes) and reservoirs, and monitoring of coastal waters such as saline estuaries is very limited (see also Chapter 3). However, lakes and reservoirs are undeniably important drinking water sources since about 82 million people in the United States (about 37 percent of the population served by public supply) rely on these surface water sources (William Wilber, USGS, personal communication, 2001). The exclusion of lakes, reservoirs, and coastal waters was a decision made early in the program because of a lack of human and financial resources. Thus, the emphasis in NAWQA continues to be freshwater streams and aquifers (Gilliom et al., 2001; Hirsch et al., 1988).

NAWQA's limited water quality data collection in lakes and reservoirs is oriented toward basic status and trends themes but not understanding. In its review of the NAWQA pilot program, the National Research Council (NRC, 1990) noted that long-term trends of the water quality of lakes and estuaries should be part of a national assessment and receive the same level of attention as streams and groundwater. That report recommended initial steps to monitor lakes, including the following: (1) lakes should be included but only as they affect downstream water quality; (2) one or more of the first study units should include a lake that is a significant contributor to downstream water quality; and (3) mathematical models should be developed at the initial stages for study units involving lakes. Therefore, although lakes, reservoirs, and coastal waters (estuaries) are not currently a major focus of the program, there is (at least in theory) a provision for the program to expand in that direction.

As noted previously, there are about a dozen drinking water reservoirs that NAWQA does sample at the public water supply intakes (Timothy Miller, USGS, personal communication, 2000). Therefore, NAWQA's program coverage of lakes and reservoirs is, by design, unrepresentative. Nevertheless, since lakes and reservoirs can often be important components of the local or regional hydrologic system, water quality studies, and public drinking water supplies, NAWQA activities in Cycle I study units that have "important" lakes and reservoirs basically were limited to sampling upstream and downstream. Thus, large reservoirs and lakes are essentially treated as "black boxes." Sampling in lakes and reservoirs was generally restricted to examining sedimentation and contaminant concentrations in cores (e.g., Callender and Van Metre, 1997). While NAWQA personnel can assess status and trends in water quality in some lakes and reservoirs, they are not positioned to assess understanding. Likewise, in some study units that are tributary to estuaries, NAWQA is measuring chemical fluxes into those coastal water bodies, but it is not examining resulting estuarine water quality or processes in the estuaries.

### Importance of Lakes, Reservoirs, Coastal Waters, and Permafrost

The omission of data on lakes and reservoirs in a national water quality assessment means that a primary source of public drinking water is not fully taken into consideration. While NAWQA sampling of streams flowing into reservoirs and lakes provides some measure of contaminant occurrence in source waters, considerable processing can take place in the reservoir or lake that is not accounted for. As noted previously, 37 percent of the population served by public water supplies receives its water from lakes and reservoirs. In addition, the long-term interests of commercial and sport fishing industries, recreational activities, and wildlife protection may not be served if lakes, reservoirs, and coastal waters are not part of the national assessment. Lack of evaluation of water quality status and trends in these water bodies means that effective assessment of pollutants or ecosystem stressors is not possible. As noted by the NRC (1990), it is important to identify and understand water quality trends before they reach the crisis stage. Degradation of water resources used for drinking water, recreation, and wildlife habitat would constitute significant health, aesthetic, and financial losses.

In terms of scientific understanding and management capability, the exclusion of lakes and reservoirs precludes an opportunity to understand significant physical, chemical, and biological processes that alter water quality. For example, an essential difference between lakes or reservoirs and streams is that lakes and reservoirs have longer residence times. Stream residence times are on the order of hours, days or perhaps months, whereas lakes and reservoirs have residence times on the order of months and years. Thus, lakes and reservoirs act as settling basins that retain substances as water flows through them, often resulting in an improvement of water quality in the outflows compared to the inflows. For this reason, lake sediments are often the best place to look for detection of (especially hydrophobic) toxic substances that accumulate in the aquatic environment. However, as previously noted, there has been a limited amount of lake sediment coring and analyses conducted in Cycle I of NAWQA (e.g., Callender and Van Metre, 1997). Since few lake sedimentation data are being collected, future application of integrated models that link watersheds to lakes and reservoirs may not be possible.

In addition to having longer residence times, lakes and reservoirs possess a large diversity of habitats caused by variations in temperature stratification, depth, and light penetration. This compartmentalization results in thriving communities of unique combinations of organisms that have a variety of effects on water quality. These effects range from the release of substances into the water column (e.g., internal loading of phosphorus during anoxia) to the removal of substances from the water column through processes such as sedimentation and bio-accumulation by higher organisms. The lacustrine environment amplifies the roles that many processes play in altering water quality, especially those related to sedimentation and those resulting from the development of permanent and transient biological communities. The short residence time of water in streams means that reactions and processes have less time to occur, so their effects are often more subtle and more difficult to measure. Omission of the measurement of water quality parameters in lakes and reservoirs reduces the ability to detect the effects of biological processes.

The planned Cycle II study unit reduction to 42 still appears to maintain good coverage of systems in which lakes and reservoirs are important. Watersheds with lakes and reservoirs in a variety of hydrologic settings still exist in the reduced Cycle II coverage, for example, Connecticut, Housatonic, and Thames River Basins (CONN); Mobile River Basin (MOBL); Rio Grande Valley (RIOG); Sacramento River Basin (SACR); Tennessee River Basin (TENN); Trinity River Basin (TRIN); and Willamette Basin (WILL). In most of these planned study units, lakes and reservoirs have multipurpose roles, providing water for uses other than drinking water. No weight is given to the trophic condition of the lakes and reservoirs, however, so it is unknown how many or what percentage are eutrophic, oligotrophic, and so forth.

Streams and groundwaters that are sources to coastal waters and estuaries are still well represented relative to the original (planned) 60 study units. In fact, the reduction to 42 study units does not concede much in the way of the original coverage of watersheds draining to estuaries. Still present are the following study units, all of which have estuarine systems: Acadian-Ponchartrain Drainages (ACAD); Apalachicola-Chattahoochee-Flint River Basins (ACFB); Albemarle-Pamlico Drainages (ALBE); CONN; the Delaware River Basin (DELR); Hudson River Basin (HDSN); MOBL; Potomac River Basin and Delmarva Peninsula (PODL); SACR; and Santee River Basin and Coastal Drainages (SANT).

In addition, a case could be made that none of the proposed 42 study units includes a region of continuous or discontinuous permafrost, which would exert a strong control on the occurrence of groundwater. The lone study unit covering this regime, Cook Inlet Basin, is slated for elimination in Cycle II (Gilliom et al., 2000a). As noted earlier, this regime is not significant in terms of population and water use. However, its elimination will limit the program's ability to comment on the effects of climate change.

## Cycle II Opportunities

Despite the lack of detailed limnological and coastal water investigations in Cycle I, there are still opportunities for the Cycle II NAWQA program to assess some impacts of lakes, reservoirs, and coastal waters on water quality. The NAWQA program could compile the relative sedimentation efficiency or retention of various substances in lakes. (In this case, retention is defined as the inflow minus the outflow divided by the inflow [Dillon and Rigler, 1974].) Annual inflow and outflow loads could be used to calculate measures of retention as water moves through a lake or reservoir. For example, lakes typically retain 50 percent to 80 percent of their incoming phosphorus loads on an annual basis, and net retention is related to the trophic condition or level of biological activity (Janus, 1989). Comparison of the retention capabilities of various bodies could provide guidance as to whether it would be beneficial to examine processes in greater detail. Detailed studies could be used to target sites with low, medium, and high retention for further work, with the ultimate goal of understanding the mechanisms so that predictions, management strategies, and policy can be established.

This would be a useful contribution that might not require significant resources and might help to stimulate other studies, as well. There are also opportunities for collaboration with other agencies that are more involved with lakes, reservoirs, and coastal waters; these are discussed in more detail in Chapter 7.

## SURFACE WATER AND GROUNDWATER

As discussed previously, surface water sampling under NAWQA focuses primarily on streams rather than lakes, reservoirs, and estuaries. Nonetheless, lakes, reservoirs, and wetlands are also found in many of the Cycle I and (planned) Cycle II study units. It should be noted that although NAWQA surface water sites include a wide variety of streams and land usage (see Chapter 1), they do not represent a statistically representative sample of stream sites (Miller and Wilber, 1999).

A variety of groundwater regimes are also monitored in the NAWQA program. The hydrologic setting approach considers geologic textures (e.g., permeabilities), and the SQA ensured that the major aquifer systems are represented in the Cycle II study units. The Cycle II study units do not lose much in terms of groundwater regions: only the permafrost (Alaska) and volcanic island (Hawaii) regimes are lost from Cycle I. One could argue that since these two areas have quite unique systems that are not represented elsewhere and do not include much of the nation in terms of population and drinking water use, their exclusion does not compromise the objectives of the program.

It is interesting to note that the planned Cycle II study units represent most of the different groundwater regions of the nation as defined by Heath (1982, 1984), with the exception of Alaska, Hawaii, Puerto Rico-Virgin Islands, and the Colorado Plateau-Wyoming Basin. Heath's classification scheme is based upon five features:

1. system components and their arrangement;
2. nature of the water-bearing openings (primary or secondary origin);
3. mineral composition of the dominant aquifers (soluble or insoluble);
4. water storage and transmission characteristics of the dominant aquifers; and
5. nature and location of recharge and discharge areas.

Unlike the previously described hydrologic landscape scheme, Heath's classification scheme does account for mineral composition of the dominant aquifers—a characteristic that can influence water quality and water chemistry. Heath's scheme is more suited to groundwater studies and, unlike the hydrologic landscape scheme, avoids any hydroclimatologic criteria. Nonetheless, it provides a further indication that the Cycle II NAWQA study units are reasonably representative of the nation's groundwater regimes.

Surface water regimes are well represented in terms of wetlands, lakes, reservoirs, and the variety of streams, although they are not all extensively sampled. Ephemeral streams, which occur in a number of study units, are not specifically addressed. This potential program limitation is discussed in more detail in Chapter 5.

## CONCLUSIONS AND RECOMMENDATIONS

Cycle II of NAWQA will sample 42 study units as opposed to the 60 originally planned in Cycle I. Because of this necessary reduction, issues of representativeness and coverage are even more central to the Cycle II NAWQA program. Inferences to unsampled areas must be approached critically, since NAWQA does not use a probabilistic approach. Without a probabilistic approach, the greater number of study units is desirable. Assessing the coverage and representativeness of Cycle II is fundamental to fully address the crosscutting issues of extrapolation and aggregation raised in the committee's statement of task. This assessment provides a base for continuing to address these issues in later chapters.

However, the committee feels that the budgetary-imposed reduction in the number of study units will still maintain good coverage of the nation's water resources. The primary reason for maintaining good coverage and representativeness is the commendable and iterative use of hydrologic setting regions, coupled with the linear programming approach and expert judgment-based semiquantitative analyses to select the reduced number of study units. As a result, about 40 percent of the nation's land area and 60 percent of its drinking water use will be represented in Cycle II. The 42 study units will account for about 80 percent of the nation's agricultural land area and about 80 percent of the national pesticide and fertilizer use. Seventy-nine large metropolitan areas (>250,000 population) and approximately 76 percent of the nation's MSAs occur in the 42 Cycle II study units.

Lakes, reservoirs, and coastal waters were specifically excluded from the original NAWQA design. Nevertheless, the NAWQA program is sampling certain lakes and reservoirs that provide drinking water and is also sampling some lake and reservoir sediment. For this reason, the Cycle II emphasis on lakes and reservoirs must continue to be limited to some aspects of status and trends of water quality but cannot address understanding. Estimates are being made of chemical fluxes into certain estuaries that are extremely important to other agencies' programs (e.g., the National Oceanic and Atmospheric Administration), although NAWQA does not specifically assess coastal waters themselves.

Streams and groundwater systems are adequately covered in Cycle II despite the elimination of the Cook Inlet Basin (Alaskan) and Oahu (Hawaiian) study units. One could argue that since these two areas have quite unique systems that are not represented elsewhere and do not include much of the nation in terms of population and drinking water use, their exclusion does not compromise the objectives of the program. However, eliminating the Cook Inlet Basin Study Unit does mean that none of the proposed 42 Cycle II study units includes a region of continuous or discontinuous permafrost, thus limiting the program's ability to comment on the effects of climate change.

Certain other aspects of "representativeness" may be lacking in the Cycle II study units. For example, even though drinking water use is closely correlated with population (and thus urban areas), the use of a concept such as the urban intensity index in selecting Cycle II study units may have resulted in more "representative" (vis-à-vis urban areas) study units. In addition, it is not clear whether or not the planned Cycle II (reduced) group of study units may be representative of certain types of land use that can heavily impact water quality such as mining, the forest products industry, or the petrochemical industry, for example.

The following are the committee's recommendations:

- Maintain sampling in lakes and reservoirs that are important public water supplies, collaborating with other organizations where feasible.
- Investigate the expanded use of the LPA-SQA methodology for other aspects of the NAWQA program (e.g., identifying sampling sites in the space-for-time studies; important reservoirs and lakes).
- Clearly state the geographic extent (or other qualifiers) of any inferences or generalizations that NAWQA scientists may make. This is critical since the NAWQA title implies a study that has national statistical representativeness.
- Clarify the representativeness of NAWQA Cycle II study units related to major sources of potential pollution, such as mining, forest products, petrochemical, and related industries.

## REFERENCES

Callender, E., and P. C. Van Metre. 1997. Reservoir sediment cores show U.S. lead declines. Environmental Science and Technology 31(9):424A-428A.

Dillon, P. J., and F. H. Rigler. 1974. A test of a simple nutrient budget model predicting the phosphorus concentration in lake water. Journal of the Fisheries Research Board of Canada 31:1771-1778.

EPA (U.S. Environmental Protection Agency). 1991. Environmental Monitoring and Assessment Program Guide. EPA/620/R-93/012. Washington, D.C.: U.S. Environmental Protection Agency, Office of Research and Development.

EPA. 1997. Environmental Monitoring and Assessment (EMAP) Research Strategy. EPA/620/R-98/001. Washington, D.C.: U.S. Environmental Protection Agency, Office of Research and Development.

Gilliom, R. J., W. M. Alley, and M. E. Gurtz. 1995. Design of the National Water Quality Assessment Program—Occurrence and Distribution of Water-Quality Conditions. U.S. Geological Survey Circular 1112. Sacramento, Calif.: U.S. Geological Survey.

Gilliom, R. J., K. Bencala, W. Bryant, C. Couch, N. Dubrovsky, D. Helsel, I. James, W. Lapham, M. Sylvester, J. Stoner, W. Wilber, D. Wolock, and J. Zogorski. 2000a. Prioritization and Selection of Study Units for Cycle II of NAWQA. U.S. Geological Survey NAWQA Cycle II Implementation Team. Draft for internal review. Sacramento, Calif.: U.S. Geological Survey.

Gilliom, R. J., K. Bencala, W. Bryant, C. A. Couch, N. M. Dubrovsky, L. Franke, D. Helsel, I. James, W. W. Lapham, D. Mueller, J. Stoner, M. A. Sylvester, W. G. Wilber, D. M. Wolock, and J. Zogorski. 2000b. Study-Unit Design Guidelines for Cycle II of the National Water Quality Assessment (NAWQA). U.S. Geological Survey NAWQA Cycle II Implementation Team. Draft for internal review (11/22/2000). Sacramento, Calif.: U.S. Geological Survey.

Gilliom, R. J., P. A. Hamilton, and T. L. Miller. 2001. The National Water-Quality Assessment Program—Entering a New Decade of Investigations. U.S. Geological Survey Fact Sheet 071-01. Reston, Va.: U.S. Geological Survey. Available online at http://water.usgs.gov/pubs/FS/fs-071-041/pdf/fs07101.pdf.

Heath, R. C. 1982. Classification of ground-water systems of the United States. Ground Water 20(4):393-401.

Heath, R. C. 1984. Ground Water Regions of the United States. U.S. Geological Survey Water-Supply Paper 2242. Washington, D.C.: U.S. Government Printing Office.

Hirsch, R. M., W. M. Alley, and W. G. Wilber. 1988. Concepts for a National Water-Quality Assessment Program. U.S. Geological Survey Circular 1021. Denver, Colo.: U.S. Geological Survey.

Janus, L. L. 1989. Nutrient Residence Times in Relation to the Trophic Condition of Lakes. Ph.D. dissertation. McMaster University.

Kolpin, D. W., J. E. Barbash, and R. J. Gilliom. 1998. Occurrence of pesticides in shallow groundwater of the United States: Initial results from the national water-quality assessment program. Environmental Science and Technology 32:558-566.

Mallard, G. E., J. T. Armbruster, R. E. Broshears, E. J. Evenson, S. N. Luoma, P. J. Phillips, and K. R. Prince. 1999. Recommendations for Cycle II of the National Water Quality Assessment (NAWQA) Program. U.S. Geological Survey NAWQA Planning Team. U.S. Geological Survey Open-File Report 99-470. Reston, Va.: U.S. Geological Survey.

McMahon, G., and T. F. Cuffney. 2000. Quantifying urban intensity in drainage basins for assessing stream ecological conditions. Revised manuscript prepared for the Journal of the American Water Resources Association. Raleigh, N.C.: U.S. Geological Survey.

Messer, J. J., R. A. Linthurst, and W. S. Overton. 1991. An EPA program for monitoring ecological status and trends. Environmental Monitoring and Assessment 17:67-78.

Miller, T. L., and W. G. Wilber. 1999. Emerging drinking water contaminants: Overview and role of the National Water Quality Assessment Program. Pp. 33-42 in Identifying Future Drinking Water Contaminants. Washington, D.C.: National Academy Press.

NRC (National Research Council). 1990. A Review of the USGS National Water Quality Assessment Pilot Program. Washington, D.C.: National Academy Press.

The Nature Conservancy. 1998. Rivers of Life: Critical Watersheds for Protecting Freshwater Biodiversity. Arlington, Va.: The Nature Conservancy.

USCB (U.S. Census Bureau). 2001. About Metropolitan Areas. Available online at http://www.census.gov/population/www/estimates/aboutmetro.html.

USGS (U.S. Geological Survey). 1995. Design of the National Water-Quality Assessment Program: Occurrence and Distribution of Water-Quality Conditions. U.S. Geological Survey Circular 1112. Reston, Va.: U.S. Geological Survey.

USGS. 1999. The Quality of Our Nation's Waters: Nutrients and Pesticides. U.S. Geological Survey Circular 1225. Reston, Va.: U.S. Geological Survey.

Winter, T. C. 1995. A landscape approach to identifying environments where ground water and surface water are closely interrelated. Pp. 139-144 in Charbeneau, R. J. (ed.) Groundwater Management, Proceedings of the International Groundwater Management Symposium, San Antonio, Texas. New York, N.Y.: American Society of Civil Engineers.

Winter, T. C. 2000. The vulnerability of wetlands to climate change: A hydrologic landscape perspective. Journal of the American Water Resources Association 36(2):305-311.

Winter, T. C. 2001. The concept of hydrologic landscapes. Journal of the American Water Resources Association 37(2):335-349.

Wolock, D. M., T. C. Winter, and G. McMahon. 2000. Delineation of hydrologic setting regions in the United States using geographic information system tools and multivariate statistical analyses. Unpublished manuscript. Denver, Colo.: U.S. Geological Survey.

# 3

# NAWQA Cycle II Goals—Status

## INTRODUCTION

In Cycle II of the National Water Quality Assessment (NAWQA) Program, one of the three primary goals of the program continues to be the assessment of water quality status—that is, to provide a nationally consistent description of current water quality conditions for a large part of the nation's water resources. Reflecting a broad shift in focus from gathering occurrence and distribution data to better understanding water quality trends and cause-and-effect relationships (see Chapters 4 and 5, respectively), the status goal is slated to receive only 23 percent of available Cycle II resources (Gilliom et al., 2000b). This is a sharp decrease from 80 percent of resources dedicated to status assessments during Cycle I. (However, it should be noted that some resources designated to support water quality trend studies in Cycle II would have been considered status assessments in Cycle I.) Nonetheless, several changes in water quality status assessments have been proposed for Cycle II (see also Appendix A).

The November 2000 version of the NAWQA Cycle II Implementation Team (NIT) report describes the design and implementation strategy for Cycle II investigations (see Gilliom et al., 2000b). Much of this chapter, and indeed the entire report, are based on review of this internal NIT report (and previous drafts) and subsequent deliberations by the committee. At the time this report was written, a total of three status themes, each with two objectives (i.e., $S1$ to $S6$) were being planned for Cycle II. These themes are (1) to assess the water quality of the most important stream and groundwater resources not sampled during Cycle I; (2) to measure the concentrations and frequencies of occurrence of NAWQA target constituents in aquifers and streams used as sources of drinking water; and (3) to assess the occurrence and distribution of contaminants not yet measured by NAWQA.

The status component is the baseline for all further NAWQA activities. The various themes, objectives, and contaminants selected for monitoring in Cycle II under the status goal of NAWQA are fundamental to the priority issues selected for national synthesis. The committee was asked in its statement of task to "assess the completeness and appropriateness of priority issues (e.g., pesticides, nutrients, volatile organic compounds, trace elements) selected for broad investigation under the national synthesis component of the program." Before commenting on the proposed status themes and related objectives for Cycle II, it is appropriate to state that the committee supports these existing priority national synthesis topics—pesticides, nutrients, volatile organic compounds, and trace elements— and commends NAWQA for its past and ongoing work on these important topics. The committee also strongly supports the priority for ecological synthesis that was initiated in the last years of Cycle I; this is a very important topic to which NAWQA can make a significant contribution. In the sections that follow in this chapter, the committee further comments on additions to these issues and other proposed priorities.

In the context of assessing the water quality of resources not previously sampled, as well as the stated intent to focus on potable water sources, this chapter first explores the impact of reducing the number units for Cycle II on NAWQA's ability to achieve its stated status objectives. The continued omission of lakes, reservoirs, and coastal waters in the Cycle II program (see also Chapter 2), as it will impact NAWQA's ability to fully understand significant chemical, physical, and biological processes that affect water quality, is then explored. A significant portion of the chapter is devoted to an analysis of the last of the three status themes—that of monitoring for contaminants not previously sampled. In this regard, there are several contaminants and groups of related contaminants proposed for monitoring in Cycle II for which sampling was not conducted during Cycle I. The appropriateness of each of the proposed contaminants or contaminant groups, relative to the goals of NAWQA, is evaluated. Next, an assessment of the importance of conducting sediment monitoring as related to Cycle I activities and opportunities for Cycle II and related recommendations are provided. The chapter ends with a summary of the conclusions and recommendations of the committee concerning the status goal, themes, objectives, and corresponding investigations planned for Cycle II.

## RESOURCES NOT PREVIOUSLY SAMPLED AND DRINKING WATER SOURCES

The first two new status themes of Cycle II are to assess water resources that were not sampled during Cycle I and to focus more heavily on streams and aquifers that are sources of drinking water (Gilliom et al., 2000b). The committee concurs with this general strategy and notes that there is some similarity among the following objectives that the U.S. Geological Survey (USGS) has proposed to accomplish these status themes in that they separate out the different foci of the program:

*Objective S1:* Characterize the concentrations and distributions of NAWQA target constituents in principal aquifers and selected rivers and streams that were not included in Cycle I.

*Objective S2:* Characterize the concentrations of NAWQA target constituents in downgradient shallow groundwater and streams for (a) residential and commercial development in large metropolitan areas and (b) the most extensive agricultural settings in the nation.

*Objective S3:* Characterize the concentrations and distributions of NAWQA target constituents in aquifers and streams that have the greatest withdrawals of drinking water.

*Objective S4:* Improve the reporting and explanation of the potential risk to human health due to the presence of contaminant mixtures that are frequently found in current or potential sources of drinking water.

At present, the funds that will be devoted to accomplishing these four objectives constitute approximately 18.7 percent of the Cycle II NAWQA budget, according to the NIT report by Gilliom et al. (2000b). A detailed description of the proposed changes to stream and groundwater sampling to meet each of the objectives can be found in that report and an overview in Appendix A of this report. Because of the similarity among the objectives, many sites can potentially be chosen to satisfy multiple objectives. The sampling design will be the same for any new sites that are chosen in Cycle II compared to Cycle I (see Chapter 1 for an overview of surface and groundwater monitoring in Cycle I). According to Gilliom et al. (2000b), the types of stream sites that will be introduced are primarily new indicator sites that will provide more information on agricultural, urban, and pristine (reference) land uses. In particular, sampling sites located in the most rapidly growing urban regions of the nation were thought to be underrepresented in sampling in Cycle I. Sites located near water supply intakes will be added to address Objective *S3*.

With regard to groundwater, new sampling sites will be chosen that help complete the national trends network for groundwater and that capture the Floridan aquifer system. Other priority candidates for groundwater are shallow groundwater systems that have not yet been sampled by NAWQA and are within (1) recently urbanized portions of major metropolitan statistical areas (MSAs); or (2) important regional crop group-hydrologic landscape combinations (Winter, 1995, 2001). The regional importance of agricultural settings will be characterized based on areal extent and intensity (chemical and fertilizer use and animal density) in relation to hydrologic settings (see Gilliom et al., 2000b).

One important question is whether the 29 percent reduction in the number of monitored and planned study units from Cycle I to Cycle II (i.e., 59 to 42) will have a deleterious effect on achieving these objectives, especially *S1* to *S3*. To a certain extent, the reduction in study units has been mitigated by the linear programming approach (LPA) and semiquantitative analysis (SQA) methods that were used to select the reduced suite of study units (see Gilliom et al., 2000a, and Chapter 2 for further information on LPA and SQA). Specifically, the linear program optimized the selection of study units based on drinking water use and chose study units across a representative range of hydrologic settings. The SQA approach was a ranking designed to ensure that the most significant contaminant sources are represented, that major aquifer systems are represented, and that a broad range of aquatic biological resources were included. For these reasons, the committee feels that Objectives *S1* to *S3* are likely to be accomplished in Cycle II, despite a reduced number of study units. As noted in Chapter 2, this is important to addressing the issues of extrapolation and aggregation of data for regional and national perspectives.

Lastly, it is evident that NAWQA and the USGS have developed considerable expertise in designing and conducting water quality status assessments in the program's first decade of nationwide monitoring. Thus, the committee finds that the proposal, albeit limited in detail, to establish a national drinking water team within NAWQA as it enters Cycle II (Gilliom et al., 2000b) is both logical and appropriate. Such a team should help ensure that NAWQA staff achieve a consistent and high level of status assessments that are focused on potable water sources throughout the program.

### Lakes, Reservoirs, and Coastal Waters in Cycle II

Objective *S3* demands additional consideration of the decision made by the USGS to not monitor lakes and reservoirs under NAWQA (see also Chapter 2). In 1990, a previous National Research Council (NRC) committee suggested that long-term trends in the water quality of lakes and estuaries should receive the same level of attention as streams and groundwater. However, it did not recommend the expansion of NAWQA to include lakes and estuaries at that time because of the lack of USGS personnel and expertise in biological and chemical modeling (NRC, 1990). Instead, the report recommended that (1) lakes should be considered but only as they affect downstream water quality, (2) one or more of the first study units should include a lake that is a significant contributor to downstream water quality, and (3) mathematical models should be developed at the initial stages for study units involving lakes.

These objectives were partly met during Cycle I. Large lakes and reservoirs were essentially treated as "black boxes" with sampling done upstream and downstream of lakes and reservoirs, particularly at a dozen lakes and reservoirs that supply drinking water (Timothy Miller, USGS, personal communication, 2000).

In addition, some study units chosen for monitoring in Cycle I (e.g., the Willamette Basin and the Upper and Lower Tennessee River Basins) contain lakes and reservoirs that are hydrologically important to the watershed as a whole. However, given the change in focus and declining resources for status assessments during Cycle II, it is not clear what could be gained by initiating an extensive sampling program for lakes, reservoirs, and coastal waters such as estuaries in Cycle II (although identifying and focusing on contaminant sources to the systems, especially estuaries, is appropriate). Chapter 2 includes a discussion of the inherent limitations that result from excluding lakes, reservoirs, and coastal waters from widespread and regular monitoring in NAWQA. It should be noted, however, that most of these issues are not part of the stated objectives outlined above.

The committee reiterates here that the omission of lakes and reservoirs excludes an opportunity to understand significant natural physical, chemical, and biological processes that alter water quality, which is a priority of the "understanding" theme of Cycle II (see Chapter 5). Lakes and reservoirs are also a major source of drinking water for the nation. In addition, differential sedimentation, also known as sediment focusing, occurs prominently in lakes and reservoirs, in contrast to flowing waters, and this may amplify the "signal" of toxic substance accumulation. Thus, lake sediments may be the best place to look for early detection of hydrophobic toxic substances that accumulate in the aquatic environment. Recognizing this, NAWQA proposes to collect lake sediment cores in Cycle II (Gilliom et al., 2000b).

Despite the continued lack of investigation into the details of mechanisms and processes within lakes and reservoirs, there is still an opportunity for the NAWQA program to assess their impact on water quality in Cycle II. As mentioned above, sedimentation (and deposition of other contaminants) is one of the processes that is amplified by the presence of a lake in a watershed system. From measurements of inflow and outflow, NAWQA should compile information on the retention of sediments and other contaminants by lakes and reservoirs in Cycle II. The annual inflow loads and outflow loads as currently measured by NAWQA can be used to provide a measure of retention (i.e., including sediment burial, water column content, and atmospheric losses) as water passes through a lake or reservoir. Comparison of retention capabilities of various water bodies could provide guidance on where it would be beneficial to examine the processes at work in finer detail. Detailed studies could then be used to target sites with low, medium, and high retention for further study, with the ultimate goal of understanding the mechanisms to develop predictions, management strategies, and policy.

As far as lakes and reservoirs are concerned, the NAWQA program could maximize limited resources by partnering with and using data already collected by the U.S. Environmental Protection Agency (EPA). For example, NAWQA is already collaborating with EPA's Office of Pesticide Programs to study pesticides in lakes and reservoirs, in part to overcome existing shortcomings of the program; see also Chapter 7). The NAWQA program might benefit from collaborating with states that are conducting ambient water quality monitoring programs.

A collaborative approach also could be taken with two federal agencies that have responsibility for many of the reservoirs in the United States—the U.S. Bureau of Reclamation (USBR) and the U.S. Army Corps of Engineers (USACE). USBR and USACE reservoirs typically provide water for drinking, recreation, and irrigation, in addition to supporting aquatic life. These agencies routinely conduct water quality monitoring for a variety of uses such as aquatic life support, fish consumption, primary contact, secondary contact, drinking water supply, and agriculture. The USBR is currently assessing the quality of irrigation water supplied by its projects through its National Irrigation Water Quality Program (www.usbr.gov/niwqp/). Since irrigation return flow often empties into reservoirs (e.g., Kesterson Reservoir in the San Joaquin Valley of California and selenium poisoning of wildlife [Tanji et al., 1986]), the USBR is monitoring water quality in a number of its reservoirs. Some of the USACE districts and regions (www.usace.army.mil) have very active reservoir water quality monitoring programs. For example, the Northwestern Division of the USACE's Missouri River Region (Omaha and Kansas City Districts) includes the Water Quality Management Program-Missouri River Region Lake Projects. Specifically, the Omaha District alone is conducting water quality studies in more than 30 lakes and reservoirs. NAWQA program personnel might be able to use these data to help assess Cycle II status, trends, and understanding goals and related themes and objectives for water quality in selected study units.

The NAWQA program has been collaborating with the National Oceanic and Atmospheric Administration (NOAA) in estuarine research (see Chapter 7 for further information). In its study units that are tributary to estuaries, NAWQA has been measuring inflows and chemical fluxes. A recent study by Bricker et al. (1999) used NAWQA data to establish nutrient loadings to a number of the nation's estuaries. In addition, the SPARROW (Spatially Referenced Regressions on Watershed Attributes) model (Smith et al., 1997) was used to provide first-order estimates of nutrient loads for the year 1987 from five major sources: fertilizer, livestock wastes, point sources, atmospheric deposition, and non-agricultural sources (Bricker et al., 1999). Preston and Brakebill (1999) reported on the application of SPARROW to assess nitrogen loading in the Chesapeake Bay watershed. These are just a few examples of how the Cycle II NAWQA program can extend itself to study water quality issues in the nation's coastal waters and estuaries by partnering with other agencies such as NOAA.

*Recommendations*

- Given the reduction in resources that necessitated a smaller suite of Cycle II study units, more emphasis on sampling in lakes, reservoirs, and estuaries than already planned for Cycle II is not feasible. Hence, opportunities for partnering with other agencies should be sought. For example, NAWQA study units should focus on collaborating with and using data collected by other organizations, such as state agencies, EPA, USBR, USACE, and NOAA, in assessing the importance of pollutant loadings and surface water-groundwater interactions related to water quality in lakes, reservoirs, and estuaries. In this manner, a better characterization of their water quality and some degree of understanding of the relevant processes may be obtained with a minimum of NAWQA expenditures.

- Current sampling in lakes and reservoirs that are important public supply sources should be maintained in Cycle II; other important lake-reservoir public supply sources should be included if resources become available (this might involve a reassessment of which lakes or reservoirs to sample). As noted in Chapter 2, the LPA-SQA methodology could be used to help determine which lakes and reservoirs to study.

- From measurements of inflow and outflow, NAWQA should compile information on the retention of sediments and other contaminants by lakes and reservoirs.

## Human Health Risks

Objective *S4* of the Cycle II design guidelines, as currently stated, represents a significant departure from the USGS's traditional areas of focus because it ventures into human health risk assessment. The committee emphasizes that the strength of NAWQA is in its ability to design sound sampling strategies; to use standardized, proven analytical methodologies to analyze those samples, and to evaluate the data for possible trends, causative factors, and so forth. In contrast, toxicological research and health risk assessment are more appropriately the purview of other agencies such as the National Institute of Environmental Health Science and its National Toxicology Program, the National Cancer Institute, the EPA, and the Agency for Toxic Substances and Disease Registry. The committee feels strongly that the area of human health risk assessment should not be an activity of the NAWQA program. Rather, NAWQA should concentrate on providing good water quality and related data and thorough analyses of those data. In this regard, the committee notes that NAWQA did an excellent job of synthesizing water chemistry data and biological data to assess associations between contaminant occurrence and evidence of potential endocrine disruption in fish (Goodbred et al., 1996). In conjunction with other local, state, and federal agencies, further assessments such as this might be accomplished in Cycle II, with NAWQA helping to further define hypotheses that should be toxicologically evaluated.

*Recommendation*

- NAWQA should consider significantly revising the language of Objective *S4* to fit its strengths, for example: "Describe the occurrence and co-occurrence of contaminants and contaminant mixtures in the environment that should be considered by other agencies for toxicological and health effects research and risk assessments."

## CONTAMINANTS NOT PREVIOUSLY SAMPLED

Although there will be a major decline in the proposed budget for status assessments in Cycle II versus those conducted in Cycle I, it has been proposed that NAWQA expand to include a third new status theme for contaminants that have become high national priorities in the last decade (Gilliom et al., 2000b). With respect to contaminants not previously sampled, NAWQA proposes the following two objectives:

*Objective S5:* Characterize the frequencies of occurrence and concentrations of emerging contaminants in streams and aquifers that are sources of drinking water and in streams representative of potential ecological effects from urban and agricultural land uses.

*Objective S6:* Characterize the concentrations and distributions of total and methyl mercury in streams that have the greatest potential for human exposure to mercury through consumption of drinking water or fish.

A number of potential candidates have been suggested for monitoring in Cycle II study units to include methyl mercury, waterborne microbial pathogens, new pesticides, pharmaceutical products, and high-production-volume industrial chemicals (Gilliom et al., 2000b). The approach suggested to accomplish this objective is to conduct a pilot evaluation of selected contaminants that are of high national priority, for which there are established and reliable analytical methods in a small subset of the study units.

Because of the limited funds available for such a program, it is logical to limit the monitoring to those contaminants for which analytical methods exist. However, the very nature of monitoring for emerging contaminants almost always requires the development of new analytical methods for sampling in the field and/or laboratory analyses. Thus, the USGS's desire to monitor only those emerging contaminants for which there are established analytical methods presents a somewhat difficult situation.

When selecting contaminants to monitor, consideration should be given to the status of existing and planned monitoring programs being conducted by other entities. In the early, exploratory stages of monitoring for a new contaminant, it may be prudent to defer to other organizations that have greater resources and expertise to conduct developmental work on specific methods. In this regard, NAWQA should focus on filling gaps in knowledge (e.g., by providing information in previously unstudied areas or by providing information on the temporal and spatial variability of the contaminant's occurrence and concentrations). The decision on which additional contaminants to study should be made with direct input from the EPA and other agencies interested in water quality, so that those most relevant to important regulatory and policy issues are identified. In addition, consideration should be given to the evidence for known or suspected impacts on human or environmental health. For example, it is well documented that human pathogens in water cause thousands of cases of illness annually in the United States (e.g., Barwick et al., 2000; see more below). On the other hand, the potential impacts that trace quantities of pharmaceuticals and their degradation products in water have on human or ecological health are uncertain (Daughton and Ternes, 1999; Halling-Sorensen et al., 1998).

NAWQA personnel have already used ranking schemes to help determine which pesticides and volatile organic compounds (VOCs) should be included in the program (Miller and Wilber, 1999). However, the entire suite of constituents for which monitoring is being conducted, as well as those under consideration for monitoring, must be considered as a whole. In other words, NAWQA should develop a procedure whereby contaminants can be evaluated or ranked against one another in a manner that will allow decisions about which contaminants to include in the monitoring program to be made on an objective basis. For example, a method has to be developed whereby specific pathogens can be compared to specific pesticides so that their relative importance for monitoring purposes can be determined. Furthermore, development of such a method or procedure should involve the direct participation or input of agencies interested in water quality such as the EPA.

The intrinsic difficulty of identifying a manageable list of constituents for monitoring in Cycle II raises the question of what kind of process or method is best suited to this type of judgment. In this regard, the committee notes that a recent NRC report, *Classifying Drinking Water Contaminants for Regulatory Consideration* (NRC, 2001), recommended and convincingly demonstrated a novel approach to help EPA sort thousands of potential drinking water contaminants of all types (e.g., chemicals, microorganisms, radionuclides) into two discrete sets—one that may undergo research or monitoring of some sort preparatory to an eventual regulatory decision and another much larger set that will not. That committee considered three broad types of strategies for accomplishing such a difficult task: expert judgment, rule-based systems, and prototype classifiers. Based on its review, that committee decided that a prototype classification approach using neural networks or similar methods would seem to be an innovative and appropriate means for EPA to consider. Although this committee does not necessarily recommend that NAWQA develop such an approach to assess and cull a very large set of constituents to a much smaller set for monitoring in Cycle II study units, it does note the flexibility of the approach.

Criteria that the NAWQA team may want to consider in developing the ranking scheme may include human health effects, ecological effects, mass of the contaminant released into the environment, analytical costs, availability of established detection methods, and the contribution that the USGS can make to existing monitoring efforts. A discussion of particular groups of related high-priority contaminants that are being considered for monitoring in Cycle II is provided below.

## Recommendations

- NAWQA should focus on filling gaps in knowledge (e.g., by providing information in previously unstudied areas or by providing information on the temporal and spatial variability of the contaminant's occurrence and concentrations), rather than on new contaminants for which methods have not been developed.
- The decision about which additional contaminants to study should be made with direct input from the EPA and other agencies so that the most important contaminants from a regulatory and policy-making standpoint can be monitored.
- The NAWQA team should develop a procedure either jointly or with the direct input of EPA or other federal agencies whereby all contaminants can be evaluated and/or ranked according to a variety of criteria, including the availability of analytical methods, known or suspected health or ecological effects, and other factors.

## Pesticides, Pharmaceuticals, and High-Production-Volume Industrial Chemicals

In part to address Objective $S5$, the USGS has considered adding several new pesticides and groups of related chemicals, such as pharmaceutical products, to the NAWQA list of analytes (Gilliom et al., 2000b). Three new groups of pesticides that have high usage and can easily be added to existing analytical methods or for which reliable methods can be established are proposed for addition. First, with some improvements in methods, a number of important organophosphate insecticides and degradates will be added. Further, using newly developed and validated methods, various sulfonyl urea herbicides may be included. The latter represents an important addition because the sulfonyl ureas are a new class of pesticides whose use is rapidly expanding as many older pesticides are being discontinued in the United States. However, their environmental fate is not well understood. Lastly, NAWQA also proposes to add an immunoassay method to analyze for glyphosate. Glyphosate is one of the most widely used herbicides, and its current standard analytical method is problematic and expensive. An immunoassay method, if proven reliable in NAWQA, could provide significant benefits to various regulatory monitoring programs as well. The committee agrees that all three groups of pesticides are appropriate and warrant addition to the list of monitored contaminants in Cycle II.

In contrast, the committee strongly recommends that pharmaceuticals and their degradates should not be added to NAWQA's analytical list until reliable sampling protocols and analytical methods can be validated. As in past cooperative efforts, hopefully NAWQA can collaborate internally with the USGS's Toxic Substances Hydrology Program and/or National Research Program to help refine and establish the requisite sampling and analytical approaches, and then move into an assessment study.

Similarly, the consideration of adding high-production-volume (HPV) industrial chemicals must be considered carefully. Many important HPV contaminants are already included in NAWQA's VOCs, trace metals, and elements monitoring. Where new chemicals can be accommodated easily in existing methods they might be included. If they will necessitate extensive protocol or methods development the committee suggests that NAWQA collaborate with others for the development work. Only if a particular HPV chemical is a clear priority (e.g., has known or potential human health or environmental impacts) can NAWQA consider expending its limited resources for such development and monitoring efforts. If other agencies provide support for needed development and monitoring work, these might be accommodated in a collaborative project.

## Recommendations

- NAWQA should not add pharmaceuticals to the list of contaminants to be monitored in Cycle II until reliable sampling protocols and analytical methods can be validated.
- NAWQA should carefully consider any additions of HPV industrial chemicals to its analytical list of monitored contaminants. Similar to new pesticides, where HPV chemicals can be accommodated easily by existing methods they might be added. Like pharmaceuticals, they should not be added until protocols and methods are validated, unless there are urgent reasons (or external support) to do so.

## Methyl Mercury

Mercury is one of the most widespread contaminants in our nation's watersheds and is associated with adverse health effects in exposed populations. Bioaccumulation of mercury (in the form of methyl mercury [MeHg]) often results from trophic transfer from contaminated sediments and water (Morel et al., 1998). Forty states have issued fish consumption advisories because of MeHg contamination—more than any other substance (Brunbaugh et al., 2000). A limited bed sediment and tissue sampling effort focused on methyl mercury is planned to be conducted once during the high-intensity phase of each Cycle II study unit (Gilliom et al., 2000b). More specifically, each study unit will select 8 to 10 stream sites representing the following three conditions: (1) background sites where mercury input is thought to be low with little wetland; (2) low potential input but high wetland areas; and (3) urban sites with a range of wetland areas. In the selection and monitoring of sites, atmospheric inputs of mercury should be considered. (As noted in Chapter 7, the National Trends Network of the USGS's National Atmospheric Deposition Program has collected mercury deposition data that can be used.)

At each site, one composite bed sediment sample will be analyzed for total mercury and MeHg and another for acid-volatile sulfides (AVSs); fish fillets from a top predator will be analyzed for total mercury (Gilliom et al., 2000b). This design is based on and expands on the results of a pilot study examining mercury contamination in fish at 106 sites in 20 watersheds (Brunbaugh et al., 2000). That study concluded that a four-variable model (MeHg in water, percent wetland, pH, and AVS in sediments) best predicted bioaccumulation of mercury in fish (tissue concentration of mercury versus length). When correlations between bioaccumulation of mercury and individual variables were determined, MeHg concentration in water and in sediment, total mercury in water, and pH were significant. Correlations with total mercury and AVSs in sediment were not significant. Given these findings, it appears that MeHg concentration in water should also be determined. Because of its widespread occurrence and potential health effects, the committee believes that MeHg is an important contaminant to include in Cycle II monitoring. Furthermore, the sampling design proposed by the USGS (Gilliom et al., 2000b) is adequate, except that consideration should be given to sampling MeHg concentration in the water, since the pilot study indicates that to be a good predictor of MeHg contamination in fish.

*Recommendation*

- In the selection and monitoring of MeHg sites already planned for Cycle II, consideration should be given to atmospheric inputs of mercury. NAWQA should also consider sampling MeHg concentration in the water since that appears to be a good predictor of bioaccumulation of mercury in fish.

## Microbiological Monitoring

The mission of the Water Resources Division (WRD) of the USGS is to provide reliable, impartial information needed to understand the nation's water resources. Among other goals, the WRD actively promotes the use of this information by decision makers to protect and enhance water resources for human health, aquatic health, and environmental quality and to effectively manage groundwater and surface water resources for domestic, agricultural, commercial, industrial, recreational, and ecological uses.

If one assesses the NAWQA program's goals in the context of the mission of the WRD, it becomes apparent that it is essential that the biological component of the program be increased. To obtain information that will enable decision makers to protect water resources for human and aquatic health, it is essential to have a better understanding of the biological components of those resources. Likewise, effectively managing water resources for domestic, recreational, and ecological purposes requires data on the biological quality as well as the chemical and physical quality of surface water and groundwater. The WRD, through the NAWQA program, is in a unique position to gather these data, both temporally and spatially.

From the standpoint of human health, it is well documented that microorganisms present a significant public health risk. Up to 90 percent of all reported waterborne disease outbreaks in the United States are caused by pathogenic microorganisms, rather than chemical contaminants (Barwick et al., 2000, Craun, 1991; Herwaldt et al., 1992; Kramer et al., 1996; Levy et al., 1998; Moore et al., 1993). Since 1971, more than 570,000 people have been documented as having become ill from microorganisms in drinking water as summarized in Table 3-1. There have also been a number of outbreaks associated with recreational water in natural settings such as lakes and rivers (Table 3-2). The health cost of outbreaks caused by the waterborne protozoa *Giardia* is estimated to be between $1.2 billion and $1.5 billion per year (EPA, 1997b), while the cost of the 1993 Milwaukee *Cryptosporidium* outbreak is estimated to have exceeded $54 million (Health and Environmental Digest, 1994). The EPA (1997a) estimates the cost to drinking water facilities for improved microbial treatment to be about $20 billion over the next 20 years, with about half of that needed immediately. Obtaining information on the occurrence of microorganisms in surface water and groundwater is critical to determining ways to prevent outbreaks. In addition, understanding and explaining the major factors and processes affecting water quality (the main effort in Cycle II of NAWQA) cannot be fully addressed without considering waterborne pathogens.

Contamination of public drinking water systems (including microbial contaminants) is currently regulated at the national level by the Safe Drinking Water Act (SDWA) of 1974, which was most recently amended in 1996. Microbial contamination of drinking water supplies is addressed by several existing and proposed rules, including the Surface Water Treatment Rule (SWTR) of 1989, the Interim Enhanced Surface Water Treatment Rule (IESWTR) of 1999, the Information Collection Rule (ICR) of 1996, the Total Coliform Rule (TCR) of 1989, and the proposed Ground Water Rule. For example, the IESWTR builds on the SWTR and includes several new provisions, such as a health goal of zero occurrence for *Cryptosporidium*; a minimum 99 percent removal of *Cryptosporidium* for filtered water supplies; and sanitary surveys for all surface water systems regardless of size.

TABLE 3-1  Causative Agents of Waterborne Disease Associated with Drinking Water in the United States, 1971-1998

| Causative Agent | Outbreaks | | Illnesses | |
| --- | --- | --- | --- | --- |
| | Number | Percent of Total | Number | Percent of Total |
| Gastroenteritis, unknown cause[a] | 334 | 48.1 | 82,076 | 14.4 |
| Giardia | 125 | 18.0 | 28,657 | 5.0 |
| Chemical poisoning | 74 | 10.6 | 4,360 | 0.8 |
| Shigella | 46 | 6.6 | 9,395 | 1.6 |
| Viral gastroenteritis | 29 | 4.2 | 13,441 | 2.4 |
| Hepatitis A virus | 26 | 3.7 | 772 | 0.1 |
| Campylobacter | 15 | 2.2 | 5,456 | 1.0 |
| Salmonella typhimurium | 13 | 1.9 | 2,995 | 0.5 |
| Cryptosporidium | 12 | 1.7 | 421,371 | 73.9 |
| Salmonella typhi | 5 | 0.7 | 282 | < 0.1 |
| Yersinia | 2 | 0.3 | 103 | < 0.1 |
| Toxigenic Escherichia coli | 7 | 1.0 | 1,442 | 0.3 |
| Vibrio cholera | 2 | 0.3 | 28 | < 0.1 |
| Chronic gastroenteritis | 1 | 0.1 | 72 | < 0.1 |
| Dermatitis | 1 | 0.1 | 31 | < 0.1 |
| Amebiasis | 1 | 0.1 | 4 | < 0.1 |
| Cyclospora | 1 | 0.1 | 21 | < 0.1 |
| Plesiomonas shigelloides | 1 | 0.1 | 60 | < 0.1 |
| Total | 695 | 100.0 | 570,566 | 100.0 |

[a]Microbial in origin.
SOURCE: Data from Barwick et al., 2000; Craun, 1991; Herwaldt et al., 1992; Kramer et al., 1996; Levy et al., 1998; Moore et al., 1993.

The EPA recently completed a negotiated rule-making process that led to recommendations about what it should propose under the Long-Term 2 Enhanced Surface Water Treatment Rule (LT2ESWTR). Under the LT2ESWTR, EPA intends to propose (1) initial source water monitoring requirements to determine whether additional treatment for *Cryptosporidum* would be required and (2) a second round of source water monitoring, six years after the initial assessments, to determine if initial source water quality conditions have changed to warrant additional treatment. The negotiating committee also recommended that EPA develop national water quality criteria for microbial pathogens for stream segments designated by states or tribes for drinking water use (under the Clean Water Act [CWA] authorities).

The EPA also recently established a "Beach Action Plan" as a multiyear strategy to improve the monitoring of recreational water quality and the commu-

TABLE 3-2 Causative Agents of Waterborne Disease Associated with Recreational Water (Natural Settings) in the United States, 1991-1998

| Causative Agent | Outbreaks | | Illnesses | |
| --- | --- | --- | --- | --- |
| | Number | Percent of Total | Number | Percent of Total |
| Gastroenteritis, unknown cause[a] | 10 | 19.23 | 1,101 | 30.26 |
| Naegleria fowleri | 10 | 19.23 | 10 | 0.27 |
| E. coli O157:H7 | 9 | 17.31 | 293 | 8.05 |
| Shigella sonnei | 9 | 17.31 | 1,111 | 30.53 |
| Giardia | 4 | 7.69 | 65 | 1.79 |
| Cryptosporidium | 3 | 5.77 | 429 | 11.79 |
| Dermatitis | 3 | 5.77 | 152 | 4.18 |
| Norwalk virus | 3 | 5.77 | 103 | 2.83 |
| Leptospira | 1 | 1.92 | 375 | 10.31 |
| Shigella flexneri | 1 | 1.92 | 35 | 0.96 |
| Total | 52 | 100.00 | 3,639 | 100.00 |

[a]Microbial in origin.
SOURCE: Data from Barwick et al., 2000; Kramer et al., 1996; Levy et al., 1998; Moore et al., 1993.

nication of public health risks associated with pathogen-contaminated recreational rivers, lakes, and ocean beaches (EPA, 1999). Furthermore, in October 2000, the Beaches Environmental Assessment and Coastal Health Act of 2000, or "BEACH" act, was signed into effect. Among other requirements, it amends the Clean Water Act to require ocean, bay, and Great Lakes states to comprehensively test recreational beach waters for waterborne pathogens and to notify the public when contamination levels make beach water unsafe for recreational uses.

As evident from Tables 3-1 and 3-2, a number of different microorganisms have been documented to cause waterborne disease outbreaks. It is not practical to monitor all of the microorganisms that could potentially be in water. Thus, some method for determining which microorganisms would act as signals of a potential health risk must be devised. For several decades (as reflected by early drinking water regulations such as the TCR), the drinking water community has relied on coliform bacteria as indicators of the microbiological quality of water. In theory, if coliform bacteria are present pathogenic microorganisms may be present and the appropriate precautionary measures must be taken to protect public health. Conversely, the absence of coliform bacteria is meant to signify that the water does not contain pathogenic microorganisms. Over the last several years, it has become clear that this paradigm does not always hold true—pathogens have been detected in waters that do not contain detectable concentrations of coliform bacteria, sometimes at levels that can cause a disease outbreak. Thus, it is clear that a different and more reliable method for assessing the microbiological quality of water must be developed. Indeed, the specific pathogen monitoring requirements of recent drinking water supply regulations (e.g., the IESWTR) recognize that relying on indicator organisms such as coliform bacteria may not be sufficient to protect public health.

One potential alternative to monitoring for coliform bacteria is to monitor for individual pathogens (though this is often considered prohibitively expensive). Another alternative is to monitor for alternative indicators, such as coliphages (viruses that infect bacteria) or for spores of the bacteria *Clostridium perfringens*, which are ubiquitous and survive longer than coliform bacteria in the aquatic environment. For groundwater systems, monitoring for coliphages (rather than human viruses) as indicators of fecal contamination instead of or in addition to coliform bacteria has been proposed (EPA, 2000c). Coliphages are about the same size as viruses that infect humans, in contrast to coliform bacteria, which are typically tens to hundreds of times larger than viruses. This allows bacteria to be more easily removed as water percolates through soil. In addition, coliphages have similar survival characteristics to human viruses in the environment, while coliform bacteria tend to be inactivated more rapidly.

For these and other reasons, many scientists believe that monitoring for specific pathogens, rather than continuing to rely on indicator microorganisms, is essential to accurately assess the microbiological quality of water (EPA, 2000b).

The latest Cycle II NIT report's (Gilliom et al., 2000b) recommendations for microbiological monitoring represent a significant change from earlier proposals. An earlier proposal (Francy et al., 2000a) incorporated monitoring for both indicator organisms and specific waterborne pathogens for streams and groundwater. However, the latest proposed strategy provided to the committee is to monitor Cycle II stream sampling sites located near public water supply intakes just for *Escherichia coli* to determine seasonal patterns of concentrations (William Wilber, USGS, personal communication, 2001). The groundwater monitoring program will include monitoring for *E. coli*, total coliform bacteria, and coliphages.

While *E. coli* can be an important water- and foodborne pathogen and serve as an indicator of the presence of enteric pathogens, its presence or absence in the water will provide little information about the presence or absence of other waterborne pathogens of concern. Thus, the proposed reduction is a significant concern to the committee.

The strength of the currently proposed program is that efficient use of limited resources will be made by monitoring for the indicator organisms (analyses for which are at least an order of magnitude less expensive than for individual pathogens). Temporal sampling will be conducted, which is critical for microbiological sampling. The microorganisms chosen for study are compatible with those chosen for other previous and ongoing studies, allowing for easy comparison to earlier work. The analyses of groundwater samples for coliphages in addition to indicator bacteria will provide critical information about the vulnerability of those wells to contamination by viruses of human health significance. Finally, the majority of the sampling can be performed by USGS personnel with little or no additional training (Donna Francy, USGS, personal communication, 2000).

This current proposal also has several weaknesses. First, and most significantly, is the lack of any specific pathogen (e.g., enteric viruses, *Cryptosporidium*, *Giardia*) monitoring in the program. As noted previously, the presence or absence of "indicator" organisms such as *E. coli* in water does not necessarily have a direct correlation with the presence or absence of pathogenic microorganisms in water. There is a critical need to obtain information about the co-occurrence of indicators and pathogens in a given water body at a given time. There is also a need to evaluate the ability of an indicator to indicate the vulnerability of a site to microbiological contamination. In other words, it would be useful to determine whether the presence of indicators in a well is correlated with the presence of pathogens in that well at some time—not necessarily at the same time as the indicators were found. This is especially significant for groundwater systems, where contamination may be more sporadic over time. The committee recommends that if the current proposal is used for the microbiological monitoring program, any monitoring sites that have indicator-positive samples should also be tested for the presence of specific pathogens. A subset of sites that are indicator-negative should also be tested for pathogens in a manner that will allow statistical analysis of the data.

Another potential weakness of the proposed monitoring program is that some methods development may be necessary to optimize recovery of microorganisms from the samples, which will divert resources from the program. Because the NIT report does not specify the exact methods to be used, the strains of bacteria being considered as hosts for coliphage infection, or more of the details of the methods, it is difficult to determine whether development of the methods will be a significant issue.

While the committee recognizes the limitations imposed by lack of resources, it strongly recommends that NAWQA reconsider its earlier proposal (described below) for pathogen and indicator bacteria monitoring. The impact of waterborne microorganisms on human health is well documented, and such a significant group of waterborne contaminants should not be excluded from a national monitoring program. NAWQA is well positioned to contribute significantly to the knowledge base on the occurrence of microorganisms in water, especially with respect to their temporal occurrence.

The previous recommendations for microbiological monitoring made by the USGS incorporate monitoring for both indicator organisms and pathogenic organisms (Francy et al., 2000a). The proposed strategy was to monitor all surface water sites for *E. coli* and then to perform monitoring for other indicators (coliphages and *Clostridium perfringens*) and selected pathogens (*Crypto-

*sporidium, Giardia,* and enteric viruses) at a subset of the sites. In this way, the costs of the sampling program would be significantly reduced (compared to monitoring at all sites). The hope was that a correlation between indicator presence and pathogen presence could be established at those sites at which all microorganisms are monitored, enabling an assessment of the quality at sites where only indicator sampling is performed. The same strategy was proposed for groundwater wells, where sampling for bacterial indicators (total coliform bacteria, *E. coli*, and enterococci) would be performed at all sites, and coliphage and enteric virus monitoring would be performed at a subset of the sites.

This strategy was developed based on the results of a pilot monitoring project, in which microbiological analyses were performed on samples collected from six of the NAWQA study units (Francy et al., 2000b). For surface water sites, samples were analyzed for two groups of indicator bacteria and one indicator bacterium: total coliform bacteria, fecal coliform bacteria, and *Clostridium perfringens*. Significant correlations were found between the concentrations of these indicators and the following water quality characteristics: dissolved organic carbon, several nitrogen species, total phosphorus, chloride, suspended sediment, and specific conductance. For groundwater samples, analyses were performed for the following indicators: total and fecal coliform bacteria, *Clostridium perfringens*, and coliphages. Because of technical difficulties, coliphage results were not usable. Examination of the results revealed that a significant correlation existed between the detection of total coliform bacteria and aquifer type ($p < .04$). The association between total coliform bacteria detection and land use was slightly less significant ($p < .07$). Based on this limited study, the authors concluded that a greater diversity of sites and more detailed information about the sites (such as might come with full implementation as part of NAWQA) would be needed for an adequate assessment of the factors that affect the microbiological quality of water.

The strengths of the proposed program are that resources will be maximized by monitoring for the pathogenic organisms (which cost at least an order of magnitude more than the indicator microorganism per sample) at only a subset of the sites. The committee also recommends that any groundwater monitoring sites that have indicator-positive samples should be tested for the presence of pathogens. A subset of sites that are indicator-negative should also be tested for pathogens in a manner that will allow multivariate statistical analysis of the data, such as was performed by Francy et al. (2000b).

Temporal sampling will be performed, which is critical for microbiological sampling. The microorganisms chosen for study are compatible with those chosen for previous and ongoing studies, allowing for easy comparison to earlier work. Finally, the majority of the sampling can be performed by USGS personnel with little or no additional training (Donna Francy, USGS, personal communication, 2000).

There are also some potential weaknesses of this more detailed proposal.

First, all virus assays were proposed to be performed using PCR (polymerase chain reaction) only. This method does not allow one to distinguish between infective and noninfective viruses. Second, to assess accurately the vulnerability of a groundwater well to virus contamination, it is essential that viral analyses of the water be performed. The difference in size between bacteria (2-5 µm) and viruses (25-50 nm) is so great that viruses may be able to move through porous media that would not permit bacterial transport. The survival of enteric viruses, in general, is also longer than that of enteric bacteria. Therefore, the absence of bacteria in a groundwater sample does not necessarily indicate a lack of vulnerability to contamination by fecal viruses. The committee strongly recommends that coliphage analyses be performed at all groundwater wells. Finally, some methods development may be necessary to optimize recovery of microorganisms from the samples, which will divert resources from the program. (This could be considered a strength of the program in that methods can be tested on a wide variety of sample matrices.)

*Recommendations*

- Waterborne pathogens are very important to human health, and NAWQA can potentially make significant contributions to our understanding of their occurrence and distribution in the nation's waters. To this end, NAWQA should reconsider its previously proposed microbiological sampling program that would include more extensive sampling.
- Groundwater sites that have indicator-positive samples should be tested for the presence of specific waterborne pathogens. A subset of sites that are indicator-negative should also be tested for pathogens in a manner that will allow multivariate statistical analysis of the data, as performed in the study by Francy et al. (2000b).

## IMPORTANCE OF CONDUCTING SEDIMENT MONITORING

Disturbing the soil through tillage, cultivation, construction, and other land management activities increases the rate of soil erosion. Dislocated soil particles can be carried in runoff and eventually reach surface water resources, including streams, rivers, lakes, reservoirs, and wetlands. While sediment deposition at the mouths of rivers can create valuable wetlands such as the Mississippi River Delta, suspended and deposited sediments affect the utility of water resources in a number of adverse ways. Accelerated reservoir siltation reduces the useful life of reservoirs. Sediment can clog roadside ditches and irrigation canals, and block navigation channels. By raising streambeds and burying stream-side wetlands, sediment increases the probability and severity of floods. Suspended sediment can increase the cost of water treatment for municipal and industrial water uses. Suspended sediment increases the turbidity of water, possibly affecting aquatic life and the appearance of water resources to recreationists. Sedimentation of river bottoms and wetlands can destroy habitat vital to aquatic organisms. Many toxic materials can be tightly bound to clay and silt particles that are carried into water bodies, including some nutrients, agricultural chemicals, industrial wastes, metals from mine spoils, and radionuclides (Osterkamp et al., 1998). When sediment is stored, the sorbed toxins are also stored and become available for assimilation. Research on the status and trends of sediment in water systems, the impacts of sediment on aquatic life, and the sources of sediment is important because of its many substantial economic impacts on water users.

Sediment is the largest contaminant of surface water by weight and volume (Koltun et al., 1997), and it is routinely identified by states as the leading pollution problem in rivers and streams (EPA, 1998). The most recent "305(b)" water quality reports submitted by the states to EPA (as required under the CWA) indicate that sediment is a leading pollutant in 38 percent of the surveyed rivers and streams that were found to be impaired (23 percent of all rivers and streams were surveyed, and 35 percent were found to be impaired) (EPA, 2000b). Similarly, the 1998 "303(d)" reports of impaired waters pinpoint sediment as the leading cause of water impairments in the nation, affecting more than 6,000 water bodies or stream reaches (EPA, 2000a).

## Ecological Impacts and Associated Economic Costs

Suspended sediment affects aquatic organisms both directly and indirectly. High suspended sediment loads can dramatically increase mortality rates of invertebrates and fish (Newcombe and MacDonald, 1991). Accumulation of fine sediments on body surfaces and gills adversely affects macroinvertebrates (Lemly, 1982), and reduction in the feeding efficiency of fish is observed when turbidity (suspended sediment) is elevated (Barrett et al., 1992). Lower rates of primary productivity as a consequence of turbidity-limited light penetration result in reductions in food resources supporting aquatic food webs (Waters, 1995).

The effects of excess sediments deposited in the stream channel are arguably even more profound than the effects of suspended sediments because essential feeding, breeding, and refuge habitats are altered. Accumulated sediment covering insect grazing habitat leads to increased rates of drift (Rosenberg and Wiens, 1978). Sediment deposition results in more homogeneous habitat, which is linked to reductions in species richness (Townsend et al., 1997). For example, insect diversity decreases as the mean size of bed material declines (Shields and Milhous, 1992). In addition to decreasing the food resources available to fish, excess sedimentation fills interstitial spaces that are essential for successful reproduction in many fish species (Waters, 1995). This has been demonstrated numerous times for salmonids, but many other fish also require clean gravel substrates for spawning (Etnier and Starnes, 1993). Many stream fish (especially juveniles) utilize the spaces between rocks as resting sites and as over-wintering refuges, which are eliminated as these spaces are filled with sediment (Bjornn and Reiser, 1991; Newcombe and MacDonald, 1991). Overall, excess sediment can reduce biodiversity of aquatic resources and may pose threats to threatened and endangered species if it occurs in critical habitats. For example, soil erosion from logging, grazing, and agriculture could pose threats to remaining salmon habitat in the Pacific Northwest, hindering recovery efforts (Aillery et al., 1996).

An indication of the damage to water resources from sediment is the value placed on reducing sediment's impacts. Ribaudo (1989b) estimated that the Conservation Reserve Program, a U.S. Department of Agriculture (USDA) program for retiring highly erodible cropland, could result in $21.4 million per year in benefits to freshwater recreational fishing from reduced sedimentation.

Other types of water-based recreation besides fishing can be affected by sediment, as demonstrated by numerous examples. Macgregor (1988) found that sedimentation at 46 Ohio State Park lakes resulted in welfare losses to boaters ranging from less than $0.01 to $11.95 per ton of sediment, with an average of $0.49 per ton. Sedimentation in recreational lakes has also been found to affect lakeside property values (Bejranonda et al., 1999). Feather and Hellerstein (1997) looked at erosion reduction on private lands in the United States from 1982 to 1992 and estimated benefits to water-based recreation of $373 million, including fishing, boating, and swimming. They also found that almost 88 percent of the benefits accrued to recreation on lakes.

Reservoir sedimentation is another consequence of soil erosion. Survey data collected by the USDA and the U.S. Department of the Interior (USDI) indicate that in the 1970s and early 1980s, sedimentation eliminated slightly more than 0.2 percent of the nation's reservoir capacity each year (Crowder, 1987). Annual economic costs, based on replacing lost capacity, were estimated to be $819 million per year (1980 dollars).

Most municipal water treatment plants must filter water before it is treated and distributed. The greater the amount of suspended sediment, the more expensive is this filtration. Annual costs to the water treatment industry from sediment were estimated to be between $458 million and $661 million in 1984 (Holmes, 1988). A study of treatment costs for a single treatment plant with a capacity of 65 million gallons per day (MGD) in an urban watershed found that reducing suspended sediment and associated water turbidity from an average of 23 NTU (nephelometric turbidity units) to 1-2 NTU would reduce capital and operations and maintenance costs by more than $2 million per year (Davis, 1999). Reducing turbidity to an average of 9 NTU would reduce costs by more than $700,000 per year.

Sedimentation in navigation channels increases the costs to shipping by increasing transit time and decreasing the amount of cargo that can be carried. The USACE incurred dredging costs of more than $500 million per year for maintaining navigation channels over the period 1992-1998 (Cecil Davison, Economic Research Service, USDA, personal communication, 2000). The Ohio Department of Natural Resources estimated direct off-site cost of removing soil erosion sediment in Ohio at $160 million per year (Bejranonda et al., 1999).

Although these are not a measure of total damages, citizens in North Carolina valued the existence of urban erosion and sediment pollution programs at between $7.1 million and $14.9 million per year (Paterson et al., 1993). Total damages from sediment due to erosion have been estimated to be between $5 billion and $17 billion per year in the United States (Ribaudo, 1989b). These estimates include damages or costs to navigation, reservoirs, recreational fishing, water treatment, water conveyance systems, and industrial and municipal water use.

## Current Information on Sediment Status and Trends.

A lack of monitoring data to evaluate control actions aimed at reducing the impacts described above has made it difficult or impossible to assess the effectiveness of erosion and sedimentation control policies, leaving open the question of whether public resources were well spent (GAO, 1990). The USDA spends tens of millions of dollars annually on conservation programs to combat soil erosion. The Conservation Reserve Program alone was estimated to provide between $2 billion and $5 billion in water quality benefits from reduced sedimentation; however, these benefits could have been higher through better targeting (Ribaudo, 1989a,b). Current estimates of soil erosion and the impacts of conservation practices on water quality are based on models, primarily the Universal Soil Loss Equation (Wischmeier and Smith, 1978). Unfortunately, there is little physical, field-based evidence to verify model estimates of erosion or its impacts on water quality. Trimble and Crosson (2000) claim that ". . . we do not seem to have a truly informed idea of how much soil erosion is occurring in this country, let alone of the processes of sediment movement and deposition." They recommend a comprehensive national system of monitoring soil erosion and downstream sediment movement, including suspended sediment and bedload.

This is not meant to imply that sediment data are never collected. In fact, such data are collected from numerous sites in North America and have been used to assess national trends in suspended sediment and to show that sediment concentrations have trended downward in some regions (e.g., Smith et al., 1993). However, these data are deficient in several ways. Sampling sites are operated for a variety of purposes and do not represent a sampling network from which regional or national inferences about suspended sediment can be made (e.g., see Osterkamp et al., 1998; Parker et al., 1997). Federal funding is not available to measure bedload, which may account for one-half or more of the total sediment load in some streams (Osterkamp et al., 1998). Also, most sediment sampling does not include sorbed chemical loads.

From a purely scientific standpoint, adequate sediment monitoring data in streams are a prerequisite to understanding the associated chemical and biological systems. Increasingly sophisticated studies of the chemistry and ecology of river systems have placed additional demands on our understanding of the physical system of rivers, which includes sediment transport processes. However, several recent federal water quality protection efforts with practical applications would also benefit from improved sediment sampling data, such as the EPA's Total Maximum Daily Load (TMDL) Program (discussed later). There is an urgent need among state and federal agencies to evaluate the magnitude of adverse impacts on designated beneficial uses of a stream reach related to sediment and to develop appropriate actions to mitigate these impacts.

## Sediment Monitoring and Assessment in Cycle I and Cycle II of NAWQA

Occurrence and distribution assessment of water quality constituents was the major NAWQA Cycle I activity, with suspended sediment identified as one of the water quality issues to be addressed (Gilliom et al., 1995). The approach taken by NAWQA to assess the water quality of streams is based on three interrelated components: water column, bed sediment and tissue, and ecological studies (Gilliom et al., 1995). Sediment sampling is an integral part of both the water column and the bed sediment studies, and substrate conditions (i.e., stream bottom sediment) are an integral part of ecological habitat assessment. Suspended sediment is a targeted characteristic at all basic fixed sampling sites and intensive fixed sampling sites. Bed sediment is a targeted characteristic at all basic fixed sites, intensive fixed sites, and bed sediment sites. The sampling data collected on sediment are adequate to characterize suspended sediment in streams in each study unit. Bed sediment sampling is conducted primarily to measure trace elements and hydrophobic organic contaminants (Shelton and Capel, 1994). Site selection for bed sediment sampling considers factors such as depositional zones for fine-grained particles and wadability (Shelton and Capel, 1994). Such a sampling strategy is probably inadequate for measuring bed sediment loads.

The study units differed in their reporting of findings related to suspended sediment, even though all collected sediment data. Suspended sediment is not one of the seven water quality components being compared among study units (i.e., national synthesis topics), so whether a study unit reported sediment findings depended on whether this is a major local issue. In the first 36 summary reports released by the Cycle I study units, only 8 indicated that sediment was a major local issue (Central Columbia Plateau, Albemarle-Pamlico Drainages, Willamette Basin, Red River of the North Basin, Upper Snake River Basin, Lake Erie-Lake St. Clair Drainages, Upper Mississippi River Basin, and Upper Colorado River Basin). Some regions where agriculture is a major land use, such as the Eastern Iowa Basins and Lower Illinois River Basin, did not report any findings regarding sediment in streams.

Similarly, a national synthesis of sediment in surface waters is not planned for Cycle II of NAWQA. At the start of the program, one of the first planned national synthesis reports was to be on nutrients and sediments (Leahy and Wilber,

1991). However, the sediment component was subsequently dropped as part of that national synthesis topic. It seems that this decision was primarily a resource issue. It has been suggested that the USGS has had a diminished ability to conduct sediment transport research in recent decades and that it has only limited expertise in sediments (Robert Hirsch, USGS, personal communication, 1999). Rather than diverting resources to acquire this expertise, the decision was made to focus on those areas in which USGS expertise was strongest.

In Cycle II, sediment data will continue to be collected in water column and bed sediment sampling (Gilliom et al., 2000b). A planned study of the effects of changes in agricultural management practices on trends in streams will include suspended sediment. However, suspended sediment is not included in the study of the effects of urbanization on trends in streams. Sediment is, however, a significant issue in areas undergoing rapid development.

## Assessment

The USGS is in a favorable position to provide leadership in sediment monitoring and interpretation. The subject of sediment is one that is critical to many USGS activities (Gray et al., 1997), and research on sediment-related topics is conducted by all four USGS divisions. Furthermore, the USGS is not the only agency conducting sediment research. The USBR undertakes technical studies to plan and design water resource facilities such as dams and reservoirs, to improve operation and maintenance of existing facilities, and to restore fish and wildlife health (Yang and Young, 1997). The Bureau of Land Management (BLM) is studying sediment as it relates to riparian health, abandoned mine land restoration, and salinity control (USGS-BLM, 1997). USDA's Agricultural Research Service conducts research on soil erosion from agricultural lands and the movement of sediment to water resources. It also develops management practices and strategies for protecting water resources from agricultural pollutants such as sediment. However, these agencies lack the monitoring and scope necessary to describe status and trends of sediment in water resources of the United States.

Sedimentation has significant physical and budgetary effects on the ability of the USACE to accomplish its missions of keeping rivers and harbors navigable and maintaining its reservoirs (Garrett, 1997). The Waterways Experiment Station conducts research related to hydraulic and sedimentation engineering in rivers, streams, and reservoirs. Technical areas include alluvial channel and floodplain development, integrated river basin management for stabilization and restoration, and improved hydraulic design methods. The work is accomplished with the aid of numerical and physical models and field investigations. The USACE also develops equipment and procedures for collecting and analyzing sediment in conjunction with other agencies, including the USGS, as part of the Federal Interagency Sedimentation Project. The USGS participates in this project and helps organize a periodic Federal Interagency Sedimentation Conference that has been held seven times since 1947.

## Conclusion

The USGS can do more to expand the knowledge base of the impacts of sediment on water quality and on the relationships between land use, hydrology, and sediment loadings. Sediment is identified by states as the leading source of impairment in streams and rivers. NAWQA continues to collect data on suspended sediment in all of its Cycle I study units, but data are not being synthesized so that regional or national implications can be made. Programs for reducing soil erosion from agriculture and urban or suburban development could benefit by a better understanding of the linkages between land use and sedimentation and of where sedimentation is impairing aquatic health.

## Monitoring for Particle-Associated Contaminants

During Cycle I, NAWQA sampled for particle-associated constituents (e.g., trace elements, organochlorine compounds, polycyclic aromatic hydrocarbons) in all study units by sampling streambed sediments, whole fish, and bivalve soft tissues (usually *Corbicula*). A synthesis of the pesticide and polychlorinated biphenyl (PCB) component of this research in the first 20 Cycle I study units has recently been completed (Wong et al., 2000). This study presents a valuable overview of the occurrence of these compounds and their relationship to patterns of land use in the watershed. This sampling strategy will not be continued during Cycle II because several deficiencies were determined that limited its use in detecting trends, such as high variability in concentrations in bed sediments at a given site; the fact that both sediments and fish move in a stream and hence their geographic history is unknown; and an absence of target fish or bivalve species at study sites. Instead, a paleolimnological approach will be used to better detect trends in particle-associated constituents during Cycle II (Gilliom et al., 2000b). One to three sediment cores will be taken from 66 lakes or reservoirs chosen as basin integrators or as indicators of urban, agricultural, or reference conditions. Concentrations at several depths in the dated core will be used to assess trends in particle-associated constituents. The committee feels that this is a wise approach for detecting trends in levels of contaminants in sediments and linking those trends with watershed conditions. However, it provides no information on trends in contamination of aquatic biota eaten by humans. Other programs with greater responsibility for human health will have to take responsibility for monitoring and analyzing such trends.

### Recommendations

- Develop a monitoring program in each Cycle II study unit to provide information on fluxes of sediment and sorbed pollutant discharges, identify trends in sediment and sorbed contaminant loads, and allow other more appropriate agencies to assess the hazards to humans and other biota of sediment and sorbed pollutants.
- Synthesize existing habitat survey data to quantify the extent of habitat impairment resulting from excess sedimentation in Cycle II study units, and investigate the relationship between habitat impairment and land use.
- Make sediment a national synthesis topic (i.e., summarize and synthesize findings on sediment and sediment-related pollutants) and provide insights for the nation that can be used by policy makers to maximize the benefits of state and federal conservation resources, while admitting to the shortcomings of current data collection.
- Provide summaries of all study unit sediment data, as with other contaminants. This is of particular importance for watersheds that must develop a TMDL for sediment.
- Expand USGS's internal expertise in sediment monitoring and interpretation to provide a national leadership role for this important area of water quality research, and work with other local, state, and federal agencies to identify and conduct research on important sediment-related issues.

## CONCLUSIONS AND RECOMMENDATIONS

In Cycle II, it has been proposed that NAWQA increase the focus on those waters that serve as sources of potable water. The committee concurs with this general strategy, as well as that of focusing on those sources most likely to be impacted by heavy urban and agricultural activities. The committee also finds that the proposal to establish a national drinking water team within NAWQA as it enters Cycle II is both logical and appropriate. However, the committee does not feel, given the limited resources for this program, that NAWQA should enter the area of human risk assessment, as currently proposed in Objective *S4*. NAWQA has extensive expertise in designing and conducting sampling programs, synthesizing data, and assessing relationships between land use and water quality. To extend its work into risk assessment, an often contentious and litigious topic for regulatory agencies such as EPA, would require the USGS to invest in new expertise. The committee feels strongly that it would be more prudent for NAWQA to maintain its focus on its established strengths and to collaborate as necessary and appropriate with other agencies that already have the requisite risk assessment expertise in-house.

The committee strongly supports the priority issues already selected for the national synthesis component of NAWQA, (i.e., pesticides, nutrients, volatile organic compounds, trace elements, ecological synthesis; see statement of task). Further conclusions and recommendations are summarized below related to additions of contaminants to these priorities and, in particular, the addition of sediment as a national synthesis priority.

Despite the possibility that resources are going to be limiting for the NAWQA Cycle II program, the committee is pleased that the NAWQA team proposes to add some new contaminants to its list of constituents to be monitored in Cycle II. In this regard, the committee agrees that all three groups of pesticides proposed for monitoring in Cycle II are appropriate and warranted and should be added to this priority area. However, the committee feels strongly that there is a need to develop a mechanism for prioritizing all contaminants under consideration relative to one another so that optimum use can be made of the available resources. Furthermore, NAWQA should focus on monitoring sites at which similar or equivalent data are not already being collected by other agencies and for those contaminants for which there are known adverse impacts on human health and/or the environment. Other specific recommendations related to status assessments of water quality in Cycle II include the following:

- Given the reduction in resources that necessitated a smaller suite of Cycle II study units, more emphasis on sampling in lakes, reservoirs, and coastal waters such as estuaries than already planned for Cycle II is not feasible. Hence, additional opportunities for partnering with other agencies should be sought. For example, NAWQA study units should focus on collaborating with and using data collected by other organizations, such as state agencies, EPA, USBR, USACE, and NOAA, in assessing the importance of pollutant loadings and surface water-groundwater interactions to water quality in lakes, reservoirs, and estuaries. In this manner, some degree of understanding of the relevant processes may be obtained with a minimum of NAWQA expenditures.

- Current sampling in lakes and reservoirs that are important public supply sources should be maintained, and other important lake-reservoir public supply sources should be included if resources become available (this might involve a reassessment of which lakes or reservoirs to sample). As noted in Chapter 2, the LPA-SQA methodology could be used to help determine which lakes and reservoirs to study.

- From measurements of inflow and outflow, NAWQA should compile information on the retention of sediments and other contaminants by lakes and reservoirs.

- NAWQA should consider significantly revising the language (and intent) of Objective *S4* to fit its strengths, for example: "Describe the occurrence and co-occurrence of contaminants (contaminant mixtures) in the environment that should be considered by other agencies for toxicological research and risk assessments." (Various iterations of the current objective imply the NAWQA would enter into the field of risk assessment.)

- NAWQA should focus on filling gaps in knowledge (e.g., by providing information in previously unstudied areas or by providing information on the temporal and spatial variability of a contaminant's occurrence and concentrations), rather than on new contaminants for which methods have not been developed.
- The decision about which additional contaminants to study in Cycle II should be made with direct input from EPA and other agencies so that the most important contaminants from a policy-making standpoint can be monitored.
- The NAWQA team should develop a procedure either jointly or with the direct input of EPA or other agencies whereby all contaminants can be evaluated and/or ranked according to a variety of criteria, including known or suspected health or ecological effects, mass of the contaminant released to the environment, availability of analytical methods, and other factors, as part of the decision process for inclusion of new contaminants for monitoring.
- NAWQA should not add pharmaceuticals or additional HPV industrial chemicals to the list of contaminants to be monitored until reliable sampling protocols and analytical methods can be validated.
- In the selection and monitoring of MeHg sites already planned for Cycle II, consideration should be given to atmospheric inputs of mercury. NAWQA should also consider sampling MeHg concentration in the water, since this appears to be a good predictor of bioaccumulation of mercury in fish.
- The committee strongly supports the addition of waterborne pathogens and indicator microorganisms to the list of contaminants that will be monitored in Cycle II. However, NAWQA should reconsider its previously proposed microbiological sampling program (described in Francy et al., 2000a) that includes more detailed sampling, because waterborne pathogens are of such known import to human health.
- Groundwater sites that have microbiological indicator-positive samples should be tested for the presence of specific waterborne pathogens. A subset of sites that are indicator-negative should also be tested for pathogens in a manner that will allow multivariate statistical analysis of the data, as performed in the pilot study by Francy et al. (2000b).
- NAWQA should include a monitoring program in each Cycle II study unit to provide information on fluxes of sediment and sorbed pollutant discharges, identify trends in sediment and sorbed contaminant loads, and allow other more appropriate agencies to assess the hazards to humans and other biota of sediment and sorbed pollutants. (This is of particular importance for watersheds that will require a TMDL for sediment.)
- Existing habitat survey data should be synthesized to quantify the extent of habitat impairment resulting from excess sedimentation in Cycle II study units and to investigate the relationship between habitat impairment and land use (and if possible, from Cycle I data as well).

- NAWQA should make sediment a national synthesis topic (i.e., summarize and synthesize findings on sediment and sediment-related pollutants) and provide implications for the nation that can be used by policy makers to maximize the benefits of state and federal conservation resources.
- The USGS should expand its internal expertise on sediment monitoring and interpretation to provide a national leadership role for this important area of water quality research and work with other local, state, and federal agencies to identify and conduct research on important sediment-related issues.

## REFERENCES

Aillery, M. P., P. Bertels, J. C. Cooper, M. R. Moore, S. J. Vogel, and M. Weinberg. 1996. Salmon Recovery in the Pacific Northwest: Agricultural and Other Economic Effects. AER 727. Feb. Washington, D.C.: U.S. Department of Agriculture, Economic Research Services.

Barrett, J. D., G. D. Grossman, and J. Rosenfeld. 1992. Turbidity induced changes in reactive distance in rainbow trout (*Oncorhynchus mykiss*). Transactions of the American Fisheries Society 121:437-443.

Barwick, R. S., D. A. Levy, G. F. Craun, M. J. Beach, and R. L. Calderon. 2000. Surveillance for waterborne-disease outbreaks—United States, 1997-1998. MMWR 49(SS04):1-35.

Bejranonda, S., F. J. Hitzhusen, and D. Hite. 1999. Agricultural sedimentation impacts on lakeside property values. Agricultural and Resource Economics Review 28(2): 208-218.

Bjornn, T. C., and D. W. Reiser. 1991. Habitat requirements of salmonids in streams. Pp. 83-138 in Meehan, W. R. (ed.) Influences of Forest and Rangeland Management on Salmonid Fishes and Their Habitat. American Fisheries Society Special Publication 19. Bethesda, Md.: American Fisheries Society.

Bricker, S. B., C. G. Clement, D. E. Pirhalla, S. P. Orlando, and D. R. G. Farrow. 1999. National Estuarine Eutrophication Assessment: Effects of Nutrient Enrichment in the Nation's Estuaries. Silver Spring, Md.: National Oceanic and Atmospheric Administration, National Ocean Service.

Brunbaugh, W. G., D. P. Krabbenhoft, D. R. Helsel, and J. G. Wiener. 2000. A national pilot study of mercury contamination of aquatic ecosystems along multiple gradients: Bioaccumulation in fishes. Presented at the Society of Environmental Toxicology and Chemistry meeting, Nashville, Tenn. November 12-16. Available online at http://co.water.usgs.gov/trace/pubs/setac2000_hg.

Craun, G. F. 1991. Causes of waterborne outbreaks in the United States. Water Science and Technology 24(2):17-20.

Crowder, B. M. 1987. Economic costs of reservoir sedimentation: A regional approach to estimating cropland erosion damage. Journal of Soil and Water Conservation 42(3):194-197.

Daughton, C. G., and T. A. Ternes. 1999. Pharmaceuticals and personal care products in the environment: Agents of subtle change? Environmental Health Perspectives 107(Suppl. 6):907-938.

Davis, J. A. 1999. Water Treatment Cost Implications With and Without Source Water Protection. Association of State Drinking Water Administrators. Nashville, Tenn.: Jordan Jones, & Goulding, Inc.

EPA (U.S. Environmental Protection Agency). 1997a. Drinking Water Infrastructure Needs Survey: First Report to Congress. EPA 812-R-97-001. Washington, D.C.: U.S. Environmental Protection Agency, Office of Ground Water and Drinking Water.

EPA. 1997b. National primary drinking water regulations: Interim Enhanced Surface Water Treatment Rule notice of data availability; Proposed rule. Federal Register 3:59486-59557.

EPA. 1998. National Water Quality Inventory: 1996 Report to Congress. EPA 841-R-97-008. Washington, D.C.: U.S. Environmental Protection Agency, Office of Water.

EPA. 1999. Action Plan for Beaches and Recreational Waters. EPA 600-R-98-070. Washington, D.C.: U.S. Environmental Protection Agency, Office of Research and Development and Office of Water.

EPA. 2000a. Atlas of America's Polluted Waters. EPA 840-B-00-002. Washington, D.C.: U.S. Environmental Protection Agency, Office of Water.

EPA. 2000b. The Quality of Our Nation's Water: A Summary of the National Water Quality Inventory: 1998 Report to Congress. EPA841-S-00-001. Washington, D.C.: U.S. Environmental Protection Agency, Office of Water.

EPA. 2000c. National primary drinking water regulations: Ground Water Rule; Proposed rules. EPA-815-Z-00-002. Federal Register, May 10. Washington, D.C.: U.S. Environmental Protection Agency, Office of Water.

Etnier, D. A., and W. C. Starnes. 1993. The Fishes of Tennessee. Knoxville, Tenn.: University of Tennessee Press.

Feather, P., and D. Hellerstein. 1997. Calibrating benefit function transfer to assess the Conservation Reserve Program. American Journal of Agricultural Economics 79(1):151-162.

Francy, D. S., D. N. Myers, and D. R. Helsel. 2000a. Microbiological Monitoring for U.S. Geological Survey National Water-Quality Assessment Program. U.S. Geological Survey Water-Resources Investigations Report 00-4018. Columbus, Ohio: U.S. Geological Survey.

Francy, D. S, D. R. Helsel, and R. A. Nally. 2000b. Occurrence and distribution of microbiological indicators in groundwater and stream water. Water Environment Research 72(2):152-161.

GAO (U.S. General Accounting Office). 1990. Greater EPA Leadership Needed to Reduce Nonpoint Source Pollution. GAO/RECD-91-10. Washington, D.C.: General Accounting Office.

GAO. 2000. Key EPA and State Decisions Limited by Inconsistent and Incomplete Data. GAO/RCED-00-54. Washington, D.C.: General Accounting Office.

Garrett, B. 1997. Sedimentation technologies for management of natural resources in the 21st century. Proceedings of the U.S. Geological Survey Sediment Workshop, February 4-7, Reston, Va. and Harpers Ferry, W. Va. Reston, Va.: U.S. Geological Survey.

Gilliom, R. J., W. M. Alley, and M. E. Gurtz. 1995. Design of the National Water-Quality Assessment Program: Occurrence and Distribution of Water-Quality Conditions. U.S. Geological Survey Circular 1112. Sacramento, Calif.: U.S. Geological Survey.

Gilliom, R. J., K. Bencala, W. Bryant, C. Couch, N. Dubrovsky, D. Helsel, I. James, W. Lapham, M. Sylvester, J. Stoner, W. Wilber, D. Wolock, and J. Zogorski. 2000a. Prioritization and Selection of Study Units for Cycle II of NAWQA. U.S. Geological Survey NAWQA Cycle II Implementation Team. Draft for internal review. Sacramento, Calif.: U.S. Geological Survey.

Gilliom, R. J., K. Bencala, W. Bryant, C. A. Couch, N. M. Dubrovsky, L. Franke, D. Helsel, I. James, W. W. Lapham, D. Mueller, J. Stoner, M. A. Sylvester, W. G. Wilber, D. M. Wolock, and J. Zogorski. 2000b. Study-Unit Design Guidelines for Cycle II of the National Water Quality Assessment (NAWQA). U.S. Geological Survey NAWQA Cycle II Implementation Team. Draft for internal review (11/22/2000). Sacramento, Calif.: U.S. Geological Survey.

Goodbred, L. L., R. J. Gilliom, T. S. Gross, N. P. Denslow, W. B. Bryant, and T. R. Schoeb. 1996. Reconnaissance of 17β-Estradiol, 11-Ketotestosterone, Vitellogenin, and Gonad Histopathology in Common Carp of United States Streams—Potential for Contaminant–Induced Endocrine Disruption. U.S. Geological Survey Open-File Report 96-627. Sacramento, Calif.: U.S. Geological Survey.

Gray, J. R., S. J. Williams, S. E. Finger, and J. W. Jones. 1997. Overview of USGS sediment research capability. Proceedings of the U.S. Geological Survey Sediment Workshop, February 4-7, Reston, Va., and Harpers Ferry, W. Va. Reston, Va.: U.S. Geological Survey.

Halling-Sorensen, B. S. Nors Nielsen, P. F. Lanzky, F. Ingerslev, H.C. Holten Lutzhoft, and S. E. Jorgensen. 1998. Occurrence, fate and effects of pharmaceutical substances in the environment—a review. Chemosphere 36(2):357-393.

Health and Environmental Digest. 1994. *Cryptosporidium* and public health. Health and Environment Digest 8(8):61-63.
Herwaldt, B. L., G. F. Craun, S. L. Stokes, and D. D. Juranek. 1992. Outbreaks of waterborne disease in the United States: 1989-1990. Journal of the American Water Works Association 84:129-135.
Holmes, T. 1988. The offsite impact of soil erosion on the water treatment industry. Land Economics 64(4):356-366.
Koltun, G. F., M. N. Landers, K. M. Nolan, and R. S. Parker. 1997. Sediment transport and geomorphology issues in the Water Resources Division. Proceedings of the U.S. Geological Survey Sediment Workshop, February 4-7, Reston, Va. and Harpers Ferry, W. Va. Reston, Va.: U.S. Geological Survey. Available online at http://www.usgs.gov/osw/techniques/workshop/koltun.html.
Kramer, M. H., B. L. Herwaldt, G. F. Craun, R. L. Calderon, and D. D. Juranek. 1996. Waterborne disease—1993 and 1994. Journal of the American Water Works Association 88(3):66-80.
Leahy, P. P., and W. G. Wilber. 1991. National Water-Quality Assessment Program. U.S. Geological Survey Open-File Report 91-54. Reston, Va.: U.S. Geological Survey.
Lemly, D. A. 1982. Modification of benthic insect communities in polluted streams: Combined effects of sedimentation and nutrient enrichment. Hydrobiologia 87:229-245.
Levy, D. A., M. S. Bens, G. F. Craun, R. L. Calderon, and B. L. Herwaldt. 1998. Surveillance for waterborne disease outbreaks—United States, 1995-1996. MMWR Surveillance Summaries 47(SS-5):1.
Macgregor, D. R. 1988. The Value of Lost Boater Value Use and the Cost of Dredging: Evaluation of Two Aspects of Sedimentation in Ohio's State Park Lakes. Unpublished Ph.D. dissertation. Ohio State University.
Miller, T. L., and W. G. Wilber. 1999. Emerging drinking water contaminants: Overview and role of the National Water Quality Assessment Program. Pp. 33-42 in Identifying Future Drinking Water Contaminants. Washington, D.C.: National Academy Press.
Moore, A. C., B. L. Herwaldt, G. F. Craun, R. L. Calderon, A. K. Highsmith, and D. D. Juranek. 1993. Surveillance for waterborne disease outbreaks—United States, 1991-1992. MMWR Surveillance Summary 42(5):1-22.
Morel, F. M. M., A. M. L. Kraepiel, and M. Amyot. 1998. The chemical cycle and bioaccumulation of mercury. Annual Review of Ecology and Systematics 29:543-566.
Newcombe, C. P., and D. D. MacDonald. 1991. Effects of suspended sediments on aquatic ecosystems. North American Journal of Fisheries Management 11: 72-82.
NRC (National Research Council). 1990. A Review of the USGS National Water Quality Assessment Pilot Program. Washington, D.C.: National Academy Press.
NRC. 2001. Classifying Drinking Water Contaminants for Regulatory Consideration. Washington, D.C.: National Academy Press.
Osterkamp, W. R., P. Heilman, and L. J. Lane. 1998. Economic considerations of a continental sediment-monitoring program. International Journal of Sediment Research 13(4):12-24.
Parker, R. S., J. K. Sueker, and R. W. Boulger. 1997. Suspended–sediment data in environmental studies. Proceedings of the U.S. Geological Survey Sediment Workshop, February 4-7, Reston, Va. and Harpers Ferry, W. Va. Reston, Va.: U.S. Geological Survey.
Paterson, R. G., M. I. Luger, R. J. Burby, E J. Kaiser, H. R. Malcom, and A. C. Beard. 1993. Costs and benefits of urban erosion and sediment control: The North Carolina experience. Environmental Management 17(2):167-178.
Preston, S. D., and J. W. Brakebill. 1999. Application of Spatially Referenced Regression Modeling for the Evaluation of Total Nitrogen Loading in the Chesapeake Bay Watershed. U.S. Geological Survey Water-Resources Investigations Report 99-4054. Reston, Va.: U.S. Geological Survey.
Ribaudo, M. O. 1989a. Targeting the Conservation Reserve Program to maximize water quality benefits. Land Economics 65(4):320-332.
Ribaudo, M. O. 1989b. Water Quality Benefits from the Conservation Reserve Program. AER 606. Washington, D.C.: U.S. Department of Agriculture, Economic Research Service.

Rosenberg, D. M., and A. P. Wiens. 1978. Effects of sediment addition on macrobenthic invertebrates in a northern Canadian river. Water Research 12:753-763.

Shelton, L. R., and P. D. Capel. 1994. Guidelines for Collection and Processing Samples of Streambed Sediment for Analysis of Trace Elements and Organic Contaminants for the National Water-Quality Assessment Program. U.S. Geological Survey Open-File Report 94-458. Reston Va.: U.S. Geological Survey.

Shields, F. D., Jr., and R. T. Milhous. 1992. Sediment and aquatic habitat in river systems. Journal of Hydraulic Engineering 118:669-687.

Smith, R. A., R. B. Alexander, and K. J. Lanfear. 1993. Stream water quality in the conterminous United States—Status and trends of selected indicators during the 1980's. Pp. 111-140 in National Water Summary 1990-91. U.S. Geological Survey Water Supply Paper 2400. Reston, Va.: U.S. Geological Survey.

Smith, R. A., G. E. Schwarz, and R. B. Alexander. 1997. Regional interpretation of water-quality monitoring data. Water Resources Research 33(12):2781-2798.

Tanji, K., A. Lauchli, and J. Meyer. 1986. Selenium in the San Joaquin Valley. Environment 28(6):6-11, 34-36.

Townsend, C. R., M. R. Scarsbrook, and S. Dolodec. 1997. The intermediate disturbance hypothesis, refugia, and biodiversity in streams. Limnology and Oceanography 42:938-949.

Trimble, S. T., and P. Crosson. 2000. U.S. soil erosion rates—Myth and reality. Science 289:248-250.

USGS-BLM (U.S. Geological Survey-Bureau of Land Management). 1997. Proceedings of the U.S. Geological Survey Sediment Workshop, February 4-7 Reston, Va. and Harpers Ferry, W. Va. Reston, Va.: U.S. Geological Survey.

Waters, T. F. 1995. Sediment in Streams: Sources, Biological Effects, and Control. American Fisheries Society Monograph 7. Bethesda, Md.: American Fisheries Society.

Winter, T. C. 1995. A landscape approach to identifying environments where ground water and surface water are closely interrelated. Pp. 139-144 in Charbeneau, R. J. (ed.) Groundwater Management. Proceedings of the International Groundwater Management Symposium, San Antonio, Tex. New York: American Society of Civil Engineers.

Winter, T. C. 2001. The concept of hydrologic landscapes. Journal of the American Water Resources Association 37(2):335-349.

Wischmeier, W. H., and D. D. Smith. 1978. Predicting Rainfall Erosion Losses—A Guide to Conservation Planning. AH-537. Washington, D.C.: U.S. Department of Agriculture, Soil Conservation Service.

Wong, C. S., P. D. Capel, and L. H. Nowell. 2000. Organochlorine pesticides and PCBs in stream sediment and aquatic biota—Initial results from the National Water Quality Assessment Program. Sacramento, Calif.: U.S. Geological Survey.

Yang, C. T., and C. A. Young. 1997. Major sediment issues confronting the Bureau of Reclamation and research needs towards resolving these issues. Proceedings of the U.S. Geological Survey Sediment Workshop, February 4-7, Reston, Va. and Harpers Ferry, W. Va. Reston, Va.: U.S. Geological Survey.

# 4

# NAWQA Cycle II Goals—Trends and Statistical Support for Understanding

## INTRODUCTION

The second goal of the National Water Quality Assessment (NAWQA) Program that continues into Cycle II (albeit with different emphases) is the determination of observed trends (and lack of trends) in water quality and the assessment of empirical associations between land use and water quality (largely in support of the understanding theme discussed in Chapter 5). The U.S. Geological Survey (USGS) has a long and distinguished record of trends assessment in the hydrologic sciences. Indeed, many of the techniques currently employed for the estimation of trends in hydrologic time series are derived from work by USGS scientists. From a practical standpoint, this is a major focus in Cycle II because trends are often the result of water quality changes associated with human-related activities. Thus, data and information on an observed trend in water quality become particularly useful when the trend is linked to its underlying cause. The reliable and early detection of trends is of fundamental value because it can provide information on changes in water quality that might be useful for future decision making and scientific understanding relating to the management of water quality. If trends are successfully detected in a timely fashion, along with a scientific understanding of the reason for those trends, it may be possible to implement management strategies to help reduce future degradation in water quality (and/or to promote future improvements).

As in Chapter 3, the organization and content of much of this chapter is based on a careful review and subsequent committee deliberations of several iterations of the NAWQA Cycle II Implementation Team (NIT) report *Study-Unit Design Guidelines for Cycle II of the National Water Quality Assessment*

*(NAWQA)* (Gilliom et al., 2000; see Appendix A). That report describes the design and implementation strategy for Cycle II investigations in NAWQA study units. In this chapter, the committee considers the extent to which the NAWQA Cycle II design supports the three themes and six related objectives concerning the determination of trends in water quality and the association of spatial or temporal variations in water quality with urbanization and agricultural practices. In a larger sense, this chapter serves as a bridge between the statistical inference focus on water quality status assessments discussed in Chapter 3 and the scientific inference focus on understanding factors and processes that affect water quality in Chapter 5. To provide continuity between Chapters 3 and 5, this chapter includes an examination of statistical issues in trend analysis and causal inference in nonexperimental studies.

One cannot begin a discussion of trend assessment without first describing the design of water quality networks. It is only through carefully controlled water quality monitoring that trend detection is possible. Carefully controlled water quality monitoring includes attention to numerous issues including, but not limited to, site selection, determination of monitoring locations, sampling frequency and protocols, field operations, data reporting, laboratory protocols, and so on. Next, general water quality monitoring network design issues are discussed, including the development of concise monitoring objectives, the use of statistical methods for optimal design of water quality networks, and the accurate estimation of pollutant loads. When trends are the focus of a monitoring program, other important water quality monitoring design issues include the frequency and location of sampling, the length of the data series, the possible collection of collateral information that might be used to fill in missing data or augment the data series, and trend detection methods.

A sensible approach to the design of water quality monitoring networks is essential to the detection and evaluation of water quality trends—the topic of the third section of this chapter. Ideally, a water quality data series collected for trend analysis would include samples evenly spaced in time (e.g., monthly) that exhibit temporal independence and minimal measurement error. In reality, water quality data typically exhibit nonnormal distributions, seasonality, missing values, values below detection limit, changes in analytical detection methods, changes in sampling frequency and location, and serial correlation. Thus, from a practical standpoint, effective trend detection and assessment requires methods to deal with these features of real environmental data. The field of environmental statistics has emerged over the past few decades in an effort to address these complexities. Trend detection requires a rigorous background in environmental statistics. In this regard, many fundamental contributions to the field of environmental statistics have been developed and described by USGS researchers (e.g., Helsel and Hirsch, 1992).

The issue of causal inference (or scientific understanding) based on observational data is discussed in the next section of this chapter. The conventional view is that scientific understanding results from controlled experiments, and nonexperimental (observational) studies should therefore not be the basis for causal inferences. However, many established scientific fields, such as epidemiology, routinely make use of nonexperimental data for scientific inference. Beyond that, emerging work in statistical inference holds promise in this area. The NAWQA Cycle II plan proposes two research themes with a total of four related objectives to address scientific understanding of the effects of urbanization and agricultural management practices on water quality, respectively. The proposed assessments will be conducted as space-for-time studies. In this context, "space for time" implies the use of studies on many watersheds and aquifers in space, all characterized by different levels of urbanization and agricultural management practices. Analysis across many watersheds and aquifers in space, at one time, allows for an evaluation of the effects of urbanization and agricultural management practices that normally evolve over time in an individual watershed. Equivalent long-term sampling of the impacts of urbanization or agricultural management practices would require much greater budgetary commitments and far more time to get results; hence the proposed space-for-time studies are creative and efficient alternatives to time sampling.

The final section includes a summary of conclusions and related recommendations discussed throughout this chapter.

## DESIGN OF WATER QUALITY MONITORING NETWORKS AND PROGRAMS

Water quality monitoring networks exist throughout the world, and methods for the design of such networks have been organized into textbooks such as those by Sanders et al. (1994) and Harmancioglu et al. (1999). The design of water quality monitoring networks includes all activities involved in the planning and management of sampling activities to collect and process water quality data for the purpose of obtaining information about the physical, biological and chemical properties of water. In addition to collecting water quality data, monitoring activities include subsequent procedures such as laboratory analyses, data processing, storage, and ultimately the statistical analyses to produce desired information. Interestingly, the steps involved in the design of water quality monitoring systems are nearly identical to the steps involved in any data management system. An efficient data management system is required if one wishes to maximize the overall utility of the resulting data. Data collection and dissemination are costly procedures, requiring large investments. The ultimate goal of any water quality data management system is to support decision making for water quality management. To enable sensible environmental management and policy decisions, an effective and integrated water quality management system is required. The components of such a water quality management system include the following:

- overall water quality objectives and constraints;
- design of the water quality monitoring network;
- water quality sample collection;
- laboratory analyses;
- water quality data storage, distribution, and analyses;
- water quality modeling; and
- decision making for water quality management.

Since water quality monitoring network design depends on program objectives, it is vital to identify and prioritize all monitoring objectives. Typical objectives of water quality monitoring include, but are not limited to, (1) trend monitoring, (2) biological monitoring, (3) ecological monitoring, and (4) compliance monitoring. Other common goals of water quality monitoring programs include an evaluation of the impact of pollution control efforts, a determination of the current status of water quality and the detection of causes of changes in water quality processes in both space and time. As noted previously, trend monitoring forms the basis of this chapter and is required to evaluate changing water quality conditions in a watershed or aquifer as well as the results of past and future water quality management control measures.

As noted, the stated goals of a program logically dictate the resulting monitoring network design, and this can be seen clearly in the early formulation of the NAWQA program. During the late 1970s and early 1980s, the national characterization of water quality provided by the National Stream Quality Accounting Network (NASQAN) was insufficient for national policy purposes because trends identified by the NASQAN program could not be related (even grossly) to natural or anthropogenic causes (Hooper et al., 1996; see also Chapter 7). The resulting river basin assessments were well received by local constituents; however, they lacked the national perspective that would enable them to be useful to Congress. More recent versions of both the NASQAN and the NAWQA programs reflect the need (goal) to relate long-term trends (or lack of trends) in water quality to the major factors that affect observed water quality trends and conditions.

In spite of very large expenditures on water quality monitoring networks, there still remains substantial uncertainty about many aspects of water quality regimes in lakes, rivers, and coastal waters in the United States and elsewhere. As a result, there are still important advances to be made with respect to understanding the spatial and temporal patterns of water quality. Such gains will come in part from improvements in our ability to design and implement water quality monitoring networks. According to Harmancioglu et al. (1999), there is a significant gap between information needs and the information provided by current water quality monitoring networks. These researchers argue that the adoption of integrated approaches to data management appears to be the only means by which the existing gaps in information can at least be minimized. Integrated approaches to data management involve careful planning *a priori* to ensure that the databases resulting from a monitoring program are useful for a variety of programmatic objectives. One of the most common deficiencies associated with monitoring programs is the lack of coordination between different local, state, and federal monitoring agencies. (As described in Chapter 1, this was the general condition of national monitoring programs at the initiation of NAWQA in the 1980s.) This lack of coordination can lead to redundancies and inefficiencies in data collection efforts and to databases that are useful only for limited purposes. Integrated design of data collection efforts would require all data collection efforts to coordinate their activities. Water quality monitoring networks can benefit from integrated design because environmental processes are interdependent, so it is necessary to develop monitoring and management systems that permit information transfer across environmental processes. They should also be redesigned periodically to incorporate the knowledge obtained from current monitoring networks, as was accomplished by both the NAWQA and the NASQAN programs during their most recent redesign efforts.

The Cycle II NAWQA trends network design is an excellent example of the application of an integrated approach to water quality data management. NAWQA seeks to retain the local perspective enabled by its (local and regional) river quality assessments with a consistent and integrated water quality data collection approach across basins and aquifers that enables a national synthesis of results (Gilliom et al., 2000). On the one hand, two primary but separate NAWQA trends networks were developed for Cycle II—one for streams and one for groundwater—because of the different hydrological mechanisms, temporal and spatial characteristics, and assessment requirements corresponding to these two sources of water. On the other hand, ground- and surface water networks share the objectives of (1) achieving a balanced representation of the primary hydrologic landscape, ecoregion, and land-use settings of the nation and (2) emphasizing the most important resources for drinking water and aquatic organisms.

A third trend network planned for Cycle II is for contaminants in sediment. This network will focus on particle-associated trace elements and organic contaminants and is the primary approach for addressing Objectives *T4* and *T6* (see more below and in Appendix A). The National Trend Network of NAWQA Cycle II is integrated from a national perspective because sampling networks are designed to assess long-term trends for streams, groundwater, and sediment constituents using sites distributed among a wide range of environmental settings across the nation that are sampled systematically over time to evaluate trends and change. Further evidence of the integrated nature of NAWQA is provided by the dual use of both a temporal trends network and the space-for-time approach to assessing change, which is discussed earlier. The committee recommends that NAWQA continue emphasis on an integrated approach to water quality monitoring network design that attempts to coordinate efforts among various local, state, and federal agencies in an effort to make study unit designs as efficient and cost-effective as possible.

Central to NAWQA is a water quality monitoring network; thus, there should be a clearly established protocol for its design. Hirsch et al. (1988) defined the specific objectives for the design of both surface and groundwater quality investigations, and subsequent water quality monitoring networks have been designed at the study unit level based on those objectives. Trade-offs exist between the cost of monitoring, the number of water quality gages in space and the temporal frequency of measurements. The committee was, however, unable to discern any established method for evaluating the trade-offs that exist between the cost of monitoring and the spatial and temporal frequency of different monitoring networks designed to meet various objectives. Such methodologies are now commonly used in the design of water quantity networks. For example, Moss and Tasker (1991) compared regional hydrologic methods for the design of stream-gaging networks. Such methods that attempt to maximize regional information within a limited budget and time horizon have been in use for several decades. These methods exploit regional hydrologic regression methods that relate the spatial and temporal frequency of stream discharge gage sampling to the precision with which various regional streamflow characteristics are estimated. A natural extension of this methodology would be to apply the Spatially Referenced Regressions on Watershed Attributes (SPARROW) multivariate statistical water quality model for use in the design of water quality monitoring networks similar to the way in which regional multivariate statistical models of various streamflow statistics are used in the design of stream-gaging networks. The committee recommends that NAWQA attempt to extend the Network Analysis Using Generalized Least Squares (NAUGLS) approach introduced by Moss and Tasker (1991) for use in the design of water quality monitoring networks. NAWQA should develop an adequate and generalized approach to optimize the design of water quality monitoring networks that is quantitative and can be tailored to satisfy all relevant NAWQA objectives.

There is now a significant and growing body of literature on the design of water quality monitoring networks. For example, Dixon and Chiswell (1996) reviewed 150 studies related to the design of water quality monitoring networks between 1970 and 1995. They argue that this literature is quite fractured, which led them to conclude that this subject area is not evolving in a cohesive way. They further concluded that there are too few examples of case studies that represent the types of studies most useful to the water industry. Perhaps the most difficult challenge associated with the design of water quality monitoring networks relates to the multiple and competing programmatic objectives that exist. For example, competing objectives might include meeting ambient water quality targets, assessment of trends, and regulatory monitoring. Multiple objectives are commonplace in the NAWQA program since it attempts to satisfy many different constituencies. Although Harmancioglu and Alpaslan (1992) describe a procedure for the design of water quality monitoring networks that are subject to multiple and competing objectives, such research is still in its infancy. The

NAUGLS methodology (Moss and Tasker, 1991) for the design of stream-gaging networks could be extended for use in the multiobjective design of water quality monitoring networks.

Even if one assumes a single objective for the design of a water quality monitoring network, there are still many unsolved problems. For example, one programmatic objective of NAWQA is to estimate loads of selected water quality constituents at key locations. Water quality monitoring networks have been designed heuristically, with this objective in mind, so many aspects of this problem have been solved. For example, Cohn (1995) summarizes decades of research relating to the estimation of sediment and nutrient loads in rivers using statistical methods. Nearly all such estimates of long-term loads are based on statistical relationships between stream discharge ($Q$) and water quality constituent concentrations ($C$). Relationships between $C$ and $Q$ are also exploited to extend and fill in missing observations in trend evaluations, as discussed in the next section.

Although a sizable literature exists relating to estimation of relationships between $C$ and $Q$, it does not provide much guidance on the conditions under which such relations are reliable or the physical basis for such relations. Furthermore, none of the existing methods reported by Cohn (1995) seem to account for the fact that the $C$-$Q$ relationship must be combined with a much longer record of $Q$ measurements, and the resulting long-term load estimate depends heavily on the fact that there are relatively few measurements with which to estimate the $C$-$Q$ relationship, compared to the nearly continuous record of $Q$ used to estimate the long-term load. Clarke (1990) pointed out this fact when he stated that the problem of long-term load estimation is really "just an old problem (streamflow record extension and augmentation) in a new disguise." The current literature ignores the statistical issues relating to the "record extension" aspect of the load estimation problem. Furthermore, it is not clear from existing literature what method can be used to determine the relationship between the frequency of streamflow discharge sampling and water quality constituent sampling that is necessary for reliable estimation of long-term loads. Finally, comparatively little research has been performed relating to the frequency distribution of concentrations and/or loads.

It is hoped that future NAWQA and other USGS research will address all relevant issues relating to the objective of designing water quality monitoring networks to ensure accurate estimates of long-term loads. Guidance is needed on the trade-off that exists between cost of sampling and the precision associated with resulting estimates of long-term loads associated with a wide range of water quality constituents. Cost of sampling relates to the frequency of those samples in both space and time and the availability of streamflow measurements. Accuracy of resulting long-term loads is dictated by the accuracy of the concentration-discharge relationship combined with the spatial and temporal frequency of water quality and quantity samples. Quantitative methods are needed to account for these issues. Extensive exploratory analyses are required to understand the probabilistic and mechanistic structure of water quality data.

## EVALUATION OF TRENDS IN WATER QUALITY

"The primary emphasis of Cycle II is to assess long-term trends in water quality and improve our understanding of the factors and processes that govern water quality" (Gilliom et al., 2000). It is the emphasis on an explanation and description of the major factors that affect water quality conditions and trends that distinguishes NAWQA from previous USGS national water quality programs. During the 1980s, many studies identified trends in water quality (Smith and Alexander, 1985; Smith et al., 1982, 1987). The most common approach was to use the seasonal Kendall's tau test, a nonparametric test for monotonic trend detection introduced by Hirsch and others (1982). Although water quality trends were frequently observed, it was difficult or impossible to determine whether those trends arose from climate variations, natural changes, or anthropogenic influences. Furthermore, the types of trend tests applied were sensitive to contamination and analytical laboratory methods because the null hypothesis of "no trend" could be rejected for a variety of reasons, many of which had no bearing on the experiment. Problems existed with water quality data assurance management and control programs during this period, particularly for constituents present in very low concentrations, such as dissolved trace elements. The NAWQA program was designed to help reduce these types of problems and produce more meaningful regional water quality assessments and trend detection and analysis studies (Hirsch et al., 1988). A central theme planned for Cycle II is an examination of the effects of land use, including urbanization and agriculture, on the biological and chemical aspects of ground- and surface water quality (Gilliom et al., 2000). The premise is to evaluate trends in both ground- and surface waters, corresponding to a wide range of land uses and watershed use categories. For example, NAWQA hopes to highlight the detection of water quality trends in watersheds, rivers, and aquifers that are important sources of drinking water supply. Ideally, NAWQA hopes to assess water quality trends nationally, while coincidentally relating detected trends to upstream land use changes. The first two objectives of this theme of Cycle II are the following:

*Objective T1:* Determine long-term trends and changes in the concentrations of NAWQA target constituents in (a) the most important principal aquifers used for drinking water supply and (b) recently recharged groundwater upgradient of these principal aquifers in a nationally representative range of hydrologic and land use settings.

*Objective T2:* Determine long-term trends and changes in concentrations and loads of NAWQA target constituents, and in the condition of stream ecosystems, for (a) streams representative of the primary hydrologic landscape and ecoregion settings present in the study units, (b) a diversity of important stream ecosystems present in study units, (c) streams representative of agricultural, urban, reference, transitional, or mixed land-use settings in the nation, and (d) the most important streams used for drinking water supply.

There are now hundreds of studies that have examined trends in water quality across the nation. Water quality observations usually consist of intermittent series of sparse records, so that application of classical methods for trend detection has generally produced unsatisfactory results. Rigorous statistical methods are now available for testing trends in water quality data. Many of those methods are nonparametric because such methods do not require assumptions regarding the probabilistic or stochastic structure of time series of water quality data.

There are also many statistical complications associated with testing trends in water quality data such as nonnormal distributions, seasonality, missing values, values below detection limit, changes in analytical detection methods, changes in sampling frequency and location, and serial correlation, among others. Fortunately, most of these problems have now been dealt with effectively through methodological improvements. For example, many time series data sets for streamflow and associated constituent concentrations exhibit autocorrelation or serial correlation. Autocorrelation or serial correlation is a property of a data series (e.g., daily measurements of contaminant concentration) that is indicative of persistence in behavior from one data point to the next. In hydrology, autocorrelation tends to be positive, which reflects the fact that higher-than-average measurements tend to follow higher-than-average measurements (and vice versa). For example, during dry weather, streamflow is fed primarily by groundwater; thus, each day's streamflow is related to the previous day's flow. In that situation, positive autocorrelation would mean that low daily flow would tend to follow low daily flow. This temporal-structural phenomenon can confound one's ability to detect trends that may be due to longer-term causative mechanisms. To address this problem, a variety of techniques have been recommended; failure to use one of these techniques in the presence of positive autocorrelation means that a hypothesis test for trend is more likely to result falsely in the conclusion of a significant trend (see Hamed and Rao, 1998; Hirsch and Slack, 1984).

As noted above, positive serial correlation is common in water quality time series because it often reflects unmodeled "short-term persistence" in the data. Short-term persistence in water quality measurements can arise from transient (or short-term) human, hydrologic, or climatic factors that are not modeled, or adjusted for, in the trend assessment. For example, prolonged periods of higher-than-average precipitation might result in the short-term persistence of higher concentrations, which is manifested in positive autocorrelation when the resultant time series data are analyzed. This is not to be confused with the long-term anthropogenic influences that tend to cause meaningful trends. Ideally, short-term persistence or serial correlation of the time series is removed (or modeled) prior to trend detection, similar to the way in which time series are standardized prior to distributional hypothesis testing.

Recent NAWQA reports that have attempted to evaluate trends in surface (Vecchia et al., 1997) and groundwater resources (Wagner, 1997) demonstrate that most of the complexities described above have been accounted for. Such reports have clearly demonstrated NAWQA's ability to discern trends in water quality due to land use and other modifications from the basic temporal stochastic structure of the water quality time series, such as short-term persistence. Unfortunately, one of the key statistical issues—spatial correlation among water quality time series—is still usually ignored. Douglas et al. (2000) found that when they accounted for the fact that streamflow time series are correlated in space, there was very little evidence of trends in either flood and drought observations across thousands of flow sequences in the United States. However, when spatial correlation was ignored, many of the regions of the United States seemed to exhibit significant trends. These researchers show that most previous studies that ignored the correlation among flow records led to erroneous claims regarding detection of trends in streamflow. The committee suspects that the same phenomenon is relevant when testing water quality data for trends, since both concentrations and loads for many constituents may be correlated with streamflow, which is itself correlated in space, in addition to the fact that concentrations are probably correlated in space themselves.

The committee recommends that future NAWQA research relating to detection of trends in water quality data use methods similar to those described by Douglas et al. (2000) to account for the spatial correlation of water quality time series. Without a proper accounting for the spatial correlation of the water quality time series, resultant conclusions regarding regional trend assessments are likely be flawed. If it is found that the water quality records being evaluated for evidence of trends do not exhibit significant spatial correlations, then it is unnecessary to modify trend detection approaches to account for such correlations.

Another common problem arises when the relationship between streamflow and concentration is weak and that relationship is then used to determine whether trends in water quality exist. Most water quality monitoring networks do not yield continuous measurements of concentration, yet continuous measurements of discharge are commonplace. Thus, the concentration-discharge relationship (also known as the "rating curve" approach) is often used to "fill in" or "reconstruct" a time series of concentration that is subsequently tested for trend. Misspecification of the form of the concentration-discharge relationship can lead to misspecification of resulting conclusions regarding trends. Specification of the concentration-discharge relationship has received considerable attention in the literature. Although a log-linear concentration-discharge relationship may be the most commonly used assumption in practice, many other more complex relationships have been suggested. Particularly given the wide range of constituents and their properties with which NAWQA is dealing, these relationships must be reviewed carefully.

Most recent NAWQA attempts to quantify the concentration-discharge relationship have used multivariate statistical relationships such as in the SPARROW model (Smith et al., 1997) and the models summarized by Cohn (1995) and Helsel and Hirsch (1992). The performance of multivariate statistical approaches varies greatly from site to site and between constituents. In many cases, there is not enough information in the discharge record to accurately predict the concentration. Significant improvements to the log-linear concentration-discharge relationship usually result from the inclusion of model terms that account for (1) time trends, (2) seasonality, (3) serial correlation of model residuals, (4) retransformation bias, (5) hysteresis in the concentration-discharge relationships, and (6) other explanatory variables (Cohn, 1995; Helsel and Hirsch, 1992). For example, Miller (1951) showed that adjustments to the basic rating-curve model for different seasons increased the overall model performance for sediment concentrations. Additionally, Cohn et al. (1992) showed that the explanatory power of the rating-curve model increased for nutrient loads entering Chesapeake Bay when seasonality and trend parameters were added to the model. Beyond the log-linear concentration-discharge relationship, more complex transfer function-type time series models (e.g., see Lemke, 1991) are designed to handle the type of serial structure associated with concentration and discharge measurements. The committee recommends that NAWQA place greater attention on the use of transfer function methods in its development of statistically based water quality models, because such methods properly account for the stochastic structure of time series of both water quality and its correlates.

Simple bivariate and even more sophisticated multivariate statistical models are attractive because they are straightforward to estimate and employ in practice, and confidence intervals for resulting concentration and load estimates are available. However, these statistical models can exhibit lack-of-fit problems and consequently have poor predictive capability in trend studies and other assessments. Hybrid physical and statistical models are available that employ physical dynamic and spatial relationships between concentration and discharge but are parsimonious and estimable using classical statistical methods such as regression. For example, studies by O'Connor (1976) and others have developed both a physical and a statistical basis for the spatial and temporal distribution of conservative dissolved solids in rivers. Such studies could be integrated into the statistical rating curve framework currently employed by NAWQA. Similarly, simple dynamic models of the type introduced by Duffy and Cusumano (1998) can exhibit the type of time-dependent hysteresis one observes in the relationship between streamflow and concentration.

The committee recommends that NAWQA continue to emphasize practical and efficient models of the concentration-discharge relationship, because such models have applicability to trend assessments as well as many other important societal problems relating to water quality management. The committee also recommends that NAWQA place much greater emphasis in the future on the integration of physical and statistical models of the concentration-discharge relationship in the hopes that such integrative research will lead to more credible and useful models.

Finally, NAWQA and the USGS have considerable expertise in methods of water quality trend assessment. Thus, the proposal for establishment of a National Trend Team within NAWQA as it enters Cycle II is both logical and appropriate. This will help ensure that NAWQA scientists achieve a consistent and high level of trend analysis throughout the program and, hopefully, can help to further the science of trend analysis as well.

## NONEXPERIMENTAL APPROACHES TO ENHANCE SCIENTIFIC UNDERSTANDING

Conclusions about cause and effect are sometimes controversial when based on observational data since the conventional perspective in science is that causal conclusions should be based on controlled experiments. However, science has advanced in many fields (e.g., epidemiology and the social sciences, as well as hydrology) in which controlled experimentation is often not possible. NAWQA, of course, is largely an observational program, and NAWQA Cycle II goals now include a greater emphasis on causal understanding of the major factors that affect (observed) water quality. This appears to be a reasonable direction for NAWQA to pursue, particularly when supported by evidence from controlled, experimental studies and when some of the emerging approaches (see Pearl, 2001, Spirtes et al., 2001) in probability-based causal inference are employed.

In Cycle II, NAWQA proposes to use a space-for-time strategy, that is, to use data from a cross section of study units to infer behavior in a single study unit over time (Gilliom et al., 2000). Space-for-time analyses are not uncommon because they allow investigators to examine many watersheds during one period in time, instead of examining a single watershed over a long time period. Thus, less time is required for the development of scientific inferences. However, the disadvantage is that the watershed-to-watershed and aquifer-to-aquifer variability is likely to be greater than the variability in a single watershed or aquifer over time.

One advantage of the space-for-time strategy is that the set of watersheds and aquifers involved in an analysis represents a population of watersheds and aquifers that is certainly greater in number than just the NAWQA study units. Crucial to the committee's charge (which addresses aggregation and presentation of information for regional and national inferences; see also Chapters 1 and 8) is the "representativeness" of the study units. In other words, to what degree do the study units represent the range of behaviors that determine the relationships important to understanding the themes addressed in this chapter? For example, are there important urbanization effects on water quality that are not represented in the NAWQA study units? This is a judgment call, but the burden should be on NAWQA scientists to present convincing evidence that study units are sufficiently representative so that the understanding theme conclusions have regional or national relevance.

### Combining Information to Improve Inference

The commonality of design across NAWQA study units and the long-term nature of the NAWQA program provide opportunities for pooling data and model forecasts to improve analyses for a number of problems. Pooling data or "borrowing strength" from data-rich sites to infer behavior at data-poor sites is not new to USGS water resources investigations. Indeed, the USGS has been a leader in the "regionalization" of monitoring networks; for example, it has contributed much to the work on estimation of streamflow in ungaged areas, based on a regional estimator for similar nearby sites. This type of analysis, when needed, should also be useful for NAWQA water quality assessments in Cycle II.

Beyond that, the NAWQA program provides opportunities to explore additional approaches for regionalization and combining information. For example, in the past, regional estimators in stream hydrology were focused only on inferences at the sample site locations; waters between sparsely distributed stations might be assessed by simple interpolation. As an alternative, models such as SPARROW could be used in an empirical Bayesian analysis to provide a continuous estimator of water quality characteristics as a function of a weighted average of the regional- and site-specific estimators. The development of modeling-monitoring approaches such as this has great potential; the committee recommends that the NAWQA modeling team include examination of strategies that combine information from site-specific data and regional model forecasts.

## EFFECTS OF URBANIZATION

Recent analyses of threats to stream biodiversity rank land-use changes as the most significant threat (Sala et al., 2000), and urbanization is a pervasive and rapidly expanding form of land-use change with considerable impacts on water quality and aquatic biota. More than 75 percent of the U.S. population lives in urban areas (USCB, 1995), and that proportion is expected to grow in the future. Thus, NAWQA's decision to investigate the impacts of urbanization on water quality as a theme (with two objectives) in Cycle II was a wise choice.

### Cycle II Plans and Adequacy of Approach

*Objective T3:* Determine the effects of urbanization on the concentrations and distributions of NAWQA target constituents in groundwater.

*Objective T4:* Determine the effects of urbanization on the concentrations and distributions of NAWQA target constituents in streams and watersheds and on stream ecosystems.

The NAWQA Cycle II NIT stated that the existing and planned Cycle II trend network alone will not be adequate to assess water quality trends resulting from urbanization (Gilliom et al., 2000). Although this reasoning was not explained, one presumes it is because of a limited number of sampling sites combined with a likelihood of high variability in response to urbanization. Thus, to assess the effects of urbanization on stream water quality and ecosystems, the NAWQA program has adopted the previously discussed space-for-time approach, which evaluates urbanization effects at sites downstream from areas representing a range in urban development (urban gradients). As noted earlier, the space-for-time concept arises from the fact that the effects of urbanization will be assessed in a number of streams with differing degrees of watershed urbanization, and the patterns seen over this spatial gradient are assumed to represent the patterns that would be observed over time. This is a commonly used practice in ecology, and it is appropriate here. However, two problems not yet addressed by the NIT that confound interpretation of the space-for-time gradient are the importance of (1) urban watershed management practices and (2) "legacy" effects of past land uses. For example, different management practices (especially stormwater programs) within an urban watershed can result in widely differing ecological communities between streams. Similarly, two streams with identical present-day land use in their watersheds may have vastly different ecological communities because of past land-use practices (e.g., Harding et al., 1998). Dealing with these issues requires an attention to urban watershed management practices and historical land-use patterns during the site selection and data compilation phase of the research, something not currently addressed in the NAWQA site selection criteria.

An "urban intensity index" will measure urbanization along this space-for-time gradient. During the site selection process, an initial cluster analysis is done based on natural basin characteristics (e.g., geology and soils) to identify groups of watersheds with similar natural settings; then the index is used to select sites exhibiting a range of urban intensities within these groups (McMahon and Cuffney, 2000). To develop the index for the New England Coastal Basins (NECB) Study Unit (see Figure 2-1), a total of 73 land-use, infrastructure, population, and socioeconomic variables were calculated for each candidate basin. Because 1997 population size explained the greatest proportion of variation among basins, the 43 basin characteristics chosen for inclusion in the index were those most highly correlated with the 1997 population size. The index is then calculated as a multimetric index with values ranging from 0 to 100. Index values were highly correlated with impervious surface cover, and sites were chosen that corresponded to impervious surface cover of <12 percent, from 12 to 30 percent, and a very few sites with >30 percent of impervious surface cover. This was done to maximize the ability to detect response thresholds, which are expected to occur around 12 percent impervious surface cover. It is not clear how sensitive this index is to industrial and other point sources of contaminants. It is also not clear that this approach yields a more meaningful measure of urbanization intensity than would be provided by simply evaluating the percentage of impervious surface cover. Indices are useful because they allow many related factors to be combined to produce a multivariate measure; however, interpretation of an index often results in the user transforming back to the original metrics (e.g., percentage of impervious surface cover). If the urban intensity index is little more than a measure of impervious surface, then ease of interpretation may suggest reconsideration of use of the multivariate index in favor of single indicators.

A space-for-time approach can permit inferences to be made on the water quality effects likely to occur in watersheds as urbanization increases. For each gradient study, a network of approximately 30 basins representing different intensities of urban gradients will be sampled to assess the relationship among urban intensity and the in-stream physical chemical and biological responses (Gilliom et al., 2000). Streamwater quality characteristics to be sampled in Cycle II include conductance, temperature, pH, alkalinity, dissolved oxygen, nutrients, pesticides, and ions in the water column; trace elements in bed sediments; geomorphic and habitat characteristics; stage; chlorophyll a; benthic algae, invertebrate, and fish communities; and *Escherichia coli* near public water supply intakes (see Chapter 3 for further information) (Gilliom et al., 2000). The proposed sampling of trace elements in bed sediments seems out of line with what is being proposed in other parts of the NAWQA project, where it is argued that trace elements in riverine bed sediments are too variable to be interpreted. It would make more sense and be more consistent with what is being measured in other parts of the program to sample biological tissues such as *Corbicula* if available. It is the committee's understanding that chlorophyll analyses will be done each time samples are collected for algal identification according to the protocol described in Porter et al. (1993). That is, after extraction with acetone, chlorophyll is measured on a fluorometer (Carol Couch, USGS, personal communication, 2001). This is a more nationally consistent sampling program than was done in Cycle I, which is appropriate. It was also a wise decision to change to fluorometric analysis of chlorophyll since that will make the NAWQA data more comparable to data being collected by other agencies and hence more useful for establishing nutrient criteria in streams.

Water samples will be collected twice during stable baseflow—once prior to biological sampling and once in conjunction with biological sampling. Biological samples are collected once during the season appropriate to each study unit; stage and temperature will be measured continuously for a one-year period spanning the sampling events. Clearly, this sampling regime is dictated by limited resources, but it must be acknowledged that it will miss some of the major consequences of urbanization, which are most strikingly observed during storm events.

Cycle I NAWQA studies have shown that concentrations of chemical constituents in urban streams differ the most from other land-use types during stormflows (e.g., Frick et al., 1998). Because of the limited number of samples being taken, the study will have limited power to detect changes in chemical measures of water quality. Similarly, because the algal community is so sensitive to recent storm events, the ability to detect change in this parameter along the gradient is also limited. For these reasons, habitat assessment and measures of fish and invertebrate communities appear to be more powerful measures of impact because they integrate the urbanization effects experienced over a longer period of time.

One goal of the urban gradient study is to consider impacts of changes in urban water management practices on water quality (Gilliom et al., 2000). One practice not mentioned in the list of examples, but worthy of consideration, is implementation of water conservation measures. The urban gradient study planned for Cycle II presumes that land-use data of adequate quality will be provided, although this is never mentioned in the Cycle II NIT report. In rapidly urbanizing areas, one can use only the most recently available land-use coverages in data analysis. Land-use data that are five years old can be misleading of present conditions. In addition, study units need to consider the availability of historical land-use data, which will be essential to assessing the legacy effects described above.

Although detailed design guidelines have not yet been published, metropolitan areas or groups of neighboring metropolitan areas with populations >100,000 are candidate sites for urban gradient studies (although metropolitan areas >250,000 are designated as "urban" in the trend network selection). The design guidelines will be based on pilot studies being conducted in the following study units (Gilliom et al., 2000): Mobile River Basin (MOBL); Great Salt Lake Basins (GRSL); and NECB (see Figure 2-1). An examination (June 2001) of the Internet sites for the aforementioned study units revealed that only NECB (nh.water.usgs.gov/CurrentProjects/nawqa/nawqaweb.htm) had information on its space-for-time study, but neither the MOBL (tenn.er.usgs.gov/MOBL/mobl.html) site nor the GRSL site (ut.water.usgs.gov/nawqa/) had any information on its pilot studies. (See Chapter 6 for further information on communicating NAWQA data and information through the Internet.)

Groundwater flow systems will also be involved in the space-for-time approach, although the studies will be conducted a bit differently from surface water studies. For groundwater, there will be two possible approaches: (1) paired comparisons of urban and nonurban areas, specifically, sampling in paired land-use studies (LUSs), within the same hydrologic setting, preferably within the same study unit, and/or (2) sampling and comparison of groundwater chemistry along a flow path in an aquifer to access groundwaters of different ages.

The NECB's space-for-time (urban land use gradient, or ULUG) study began monitoring in early 1999. Basins are located in Massachusetts, Connecticut (this is a University of Connecticut site outside the study unit's boundaries), New

Hampshire, and Maine. The study is designed to assess the effects of urban gradients on water quality and aquatic ecosystems through the placement of six indicator sites (selected to represent areas with relatively homogeneous land use) and participation in a pilot land-use gradient study with the NAWQA Ecological Synthesis Team. The design of NECB's fixed-site indicator sites have been integrated into this study since these sites in eastern Massachusetts represent an urbanization gradient. Other studies have been linked to the ULUG design: (1) University of Connecticut studies funded by the U.S. Environmental Protection Agency (EPA) and the National Aeronautics and Space Administration (NASA; resac.uconn.edu); (2) the NECB NAWQA mercury study; (3) NECB NAWQA lake coring studies; (4) viral indicators in the urban waters study of the Massachusetts Department of Environmental Protection; and (5) the USGS's Geologic Division Clean Water Action Plan study. There are plans to conduct a SPARROW-type analysis (e.g., Smith et al., 1997) of the ULUG data.

In addition to the aforementioned three study units, the space-for-time urban gradient approach is being used in the New Jersey portion of the Long Island-New Jersey Coastal Drainages Study Unit (LINJ; nj.usgs.gov/nawqa/linj.html). The LINJ Study Unit is one of the most heavily urbanized, with more than 12,000,000 people in about 6,000 square miles. Despite this, there are productive agricultural lands around the fringes of the metropolitan areas, as well as in the more rural parts of eastern Long Island and southern and western New Jersey. However, these rural areas are rapidly becoming urban and suburban communities, making the LINJ Study Unit an ideal place for urban gradient work.

To illustrate the LINJ work, Mark Ayers, the LINJ chief, gave a presentation to the committee in August 2000 (Mark Ayers, USGS, personal communication, 2000) summarizing the accomplishments of both surface water and groundwater space-for-time studies. Ayers' presentation indicated that concentrations of volatile organic compounds (VOCs) are related to urbanization gradients. The LINJ studies also indicated that impaired fish communities can be related to urbanization (i.e., increased urbanization leads to increased impairment). In 36 northern New Jersey streams, the study found that human population and impervious surfaces appeared to impose the strongest influence on fish, invertebrate, and algal communities. Fish communities were most affected by changes in flow, total phosphorus, impervious area, and substrate composition. Invertebrate communities were most influenced by forest and wetland area, ample base flow, impervious area, and substrate composition. Algal communities were directly affected by chemical use, increases in peak flow, and substrate composition. A management perspective was also presented. A groundwater study in the vicinity of Glassboro, New Jersey, a rapidly urbanizing area, showed that nitrate and pesticides are largely from nonpoint sources and VOCs from point sources. Several nitrate management scenarios were illustrated to predict the effect on the aquifer system to the year 2050.

The LINJ space-for-time project (and other LINJ efforts) results are already having a management impact in New Jersey; presentations by the New Jersey Department of Environmental Protection (NJDEP) (Karen Schaffer, NJDEP, personal communication, 2000) and the New Jersey Office of State Planning (NJOSP) (Jim Reilly, NJOSP, personal communication, 2000) attested to the usefulness of the LINJ data and information.

Detailed comments on the study design for these objectives are not possible at this time because the final guidelines for these studies have not been completed. However, based on the LINJ results, the space-for-time approach is a worthwhile one and should provide valid results as to the effects of urbanization on water quality in both surface water and groundwater systems. When coupled with modeling, this approach will lead to significant results that will greatly assist managers in protecting our nation's waters. In addition, the following recommendations are made:

• The NAWQA Cycle II study design has focused on current land-use conditions and their relationship to stream and groundwater attributes. As data analysis proceeds, attention must also be given to urban watershed management practices and land-use history so that land-use legacies can be incorporated.

• Habitat assessment and measures of fish and invertebrate communities appear to be more powerful measures of urban impact than proposed measures of the algal community, which are sensitive to recent storm events. Thus, if resources are limiting, the algal community analyses could be eliminated with the least loss to the program.

## RESPONSE TO AGRICULTURAL MANAGEMENT PRACTICES

Agriculture is one of the dominant land uses in the United States. More than 20 percent of land is in crop production for the United States as a whole, and in some regions this percentage is much higher (e.g., 60 percent in the "Corn Belt"). Crop production necessitates major changes to the land surface, an intensive use of chemical fertilizers and pesticides, and where irrigation is used, changes to hydrology and the water budget itself. These changes have the potential for impacting the quality of surface water and groundwater resources. For example, agriculture has been identified as the major source of sediment, nutrient, and pesticide pollutants in surface water and groundwater resources and as the single largest source of water impairments (EPA, 2000). Trend analysis using NASQAN data and other research has suggested relationships between sediment concentrations and conservation programs (Smith et al., 1993), between nitrogen concentrations and the introduction of inorganic fertilizer (Goolsby et al., 1999; Smith, et al., 1993) and between the use of pesticides and groundwater quality (Barbash et al., 1999).

A systematic study of how changes in agricultural production practices affect the water quality of a watershed may better enable resource managers to identify appropriate strategies for reducing agricultural pollution. Agriculture is a dynamic system, responding to economic forces and changes in technology. The dynamic nature of cropping and livestock systems requires complementary monitoring of watershed changes in livestock numbers, cropping patterns, and other land uses and sources of pollution. Runoff from agriculture is affected by weather, which is highly variable over time. In such a setting, long-term monitoring and assessment are required to account for all stochastic influences. Although much research has been devoted to quantifying the impact of individual management practices and cropping systems on water quality, the focus of this research has largely been at the field scale. There is a need to understand the linkage among field practices, off-site movement of pollutants, and the effect on the environment at a watershed or landscape scale as it relates to surface water quality (MSEA, 1995). There is a lack of information on which fields and landscapes may be more vulnerable to off-site movement of sediment and chemicals. Long-term monitoring of water quality and changes in production and management practices should quantify the relationships between agriculture and water quality.

Over the past 20 years, hundreds of projects and programs have provided resources to farmers for reducing agriculture-related pollution such as the Rural Clean Water Program, Water Quality Incentives Program, President's Water Quality Initiative, Environmental Quality Incentive Program, Conservation Technical Assistance, Agricultural Conservation Program, Wetland Reserve Program, and Conservation Reserve Program. Changes in management practices brought about by these programs have been documented for watersheds and for larger regions. However, the impacts on water quality of most of these efforts are largely unknown. Most programs report progress in terms of surrogate parameters, such as reduced soil erosion or chemical use, rather than in terms of water quality improvements (Swader, 1993). Without direct evidence of improving water quality, farmers and others may become indifferent to the voluntary use of practices that produce environmental gains (Ribaudo et al., 1999). This could hinder the success of conservation programs. Findings from NAWQA Cycle II research can provide information to farmers and soil conservation agents about the ability of conservation practices to protect water quality.

Water quality monitoring is often incompatible with an agricultural conservation program's time scale. A water quality baseline must be established before a project is started if impacts on water quality are to be determined. This can take a number of years. It is not often feasible to delay the start of an authorized conservation program for which funds have been appropriated to establish a baseline. The length of time needed to document water quality changes may far surpass the time frame of the conservation program itself. Improvements in water quality from farmers' efforts to reduce pollutant loadings often take years to detect and document. The links between improved management and observed changes in water quality are complex. As many as 10 consecutive years of water quality data are needed before long-term changes can be distinguished from short-term fluctuations (Smith et al., 1993). Phosphorus accumulated in bottom sediments will affect water quality long after conservation practices have dramatically reduced phosphorus loadings in runoff. Similarly, fish, insects, and other biological indicators of a healthy stream may not reach acceptable levels until many years after water quality improves and riparian habitat is restored. Aquifers may take decades to show improvements in quality after chemical management is improved.

A particular Cycle II theme to be investigated is the response of water quality to long-term changes in agricultural management practices such as tillage methods, chemical use, and cropping patterns. This theme is consistent with the original framework for the NAWQA program and will provide valuable information to the U.S. Department of Agriculture (USDA), states, and agencies responsible for conservation and other agricultural programs. In particular, states have a need for this information as they develop total maximum daily load management plans to address persistent water quality problems. Many of these plans will have to include measures to reduce agricultural pollution. The U.S. General Accounting Office (GAO) concluded in 1990 that a lack of monitoring data on the scope of nonpoint source pollution and on the effectiveness of potential solutions was restricting states' ability to deal with the problem (GAO, 1990, 2000).

### Cycle II Plans and Adequacy of Approach

Spatial studies of the effects of land-use changes on water quality will be undertaken to provide explanation of findings of the trend network (Gilliom et al., 2000). The NAWQA national trend networks for streams and groundwater will monitor trends by systematic sampling over time at carefully selected sites and study areas that represent key water resources and land uses. As pointed out in the NIT Cycle II design report, the trend networks only partially address issues related to changes in urban and agricultural areas, and are inadequate for assessing the effects of urban and agricultural land uses on aquatic biota and stream ecosystems. To address Cycle II trend themes aimed at water quality changes related to agricultural practices, a range of spatial studies are planned. These will be closely integrated with the trend network and topical studies undertaken to examine the factors that govern water quality, and presumably with the activities of the Hydrologic Systems Team (HST). Some of the issues that could be addressed by these studies include the effectiveness of best management practices, the impact of animal production on water quality, and the impact of changes in chemical use practices on water quality. Two objectives are identified under this theme:

*Objective T5:* Determine the effects of long-term changes in agricultural management practices on the concentrations and distributions of NAWQA target constituents in groundwater.

*Objective T6:* Determine the effects of long-term changes in agricultural management practices on the concentrations and distributions of NAWQA target constituents in streams and watersheds and on stream ecosystems.

To address the first objective (*T5*), a subset of Cycle I LUSs and shallow groundwater study-unit survey areas that have undergone documented major changes in agricultural management practices since Cycle I will be resampled. (See Chapter 1 for further information about Cycle I water quality monitoring and study design.) Additional work will be conducted to link the history of land use change to the quality of groundwater. Special emphasis will be given to areas in which the shallow groundwater sampled through land-use studies is recharged for deeper groundwater used for drinking supplies. The studies will be located in areas where local changes in agricultural practices have occurred over several years, where hydrogeology has been well defined, and where groundwater flow can be simulated with numerical models. Flowpath studies will be conducted to investigate the links between shallow groundwater, which is most easily contaminated, and deeper groundwater used for drinking water. Flowpath studies will also be conducted to investigate links between shallow groundwater and streams. These studies will be coordinated with targeted spatial studies for streams and related understanding themes.

A space-for-time approach will be used to meet the second objective (*T6*) under the theme, premised on the hypothesis that networks of watersheds can be identified based on gradients in agricultural factors. Spatial studies will use a design that investigates a subset of watersheds chosen from the population of watersheds present in a geographic area of interest. Focus will be on management practices that vary regionally or nationally. The NIT Cycle II design report (Gilliom et al., 2000) suggests that an initial focus of studies may be provided by linking agricultural spatial studies to topical studies relevant to Objective *U13* (nutrient enrichment; see Chapter 5) and topical studies on nutrient sources and transport.

The USGS and NAWQA are well positioned to carry out this work. NAQWA has established water quality baselines and monitoring networks in the study units and is operating at a time scale sufficient to establish relationships between production practices and water quality at a watershed scale. This research will not be able to identify the water quality benefits of individual practices, but the USDA research can provide this information. Specifically, the Agricultural Research Service (ARS) evaluates new and improved strategies for reducing water contamination from agricultural lands. ARS and its State Experiment Station partners (and state Water Resources Research Institutes) develop field practices to reduce impacts of nutrients, pesticides, and other synthetic chemicals; pathogens and other bacterial contaminants; sediments; and trace elements on surface water and groundwater. These management practices and systems are evaluated for their water quality and other environmental benefits at field, farm, watershed, and basin scales on irrigated and rain-fed cropland and grazing land.

An example of what NAWQA can do regarding this theme is provided by the Lake Erie-Lake St. Clair Drainages (LERI) Study Unit (see Figure 1-1). Retrospective and other monitoring data were coupled with land-use information to evaluate the relation between suspended sediment discharges, conservation tillage practices, and soil loss in the Maumee River Basin (Myers and Metzker, 2000). In brief, increases in conservation tillage (and decreases in soil loss) were related to decreases in suspended sediment discharge from streams. This research provides a model for other study units' undertaking Cycle II analyses of the relationships between agriculture and water quality.

The NIT Cycle II report (Gilliom et al., 2000) does not elaborate on how a landscape as diverse as agriculture's would be characterized in a study that includes long-term changes in agricultural management practices. A watershed may contain hundreds of farms and thousands of individual fields. These fields may be growing a variety of crops and be farmed using a variety of tillage, conservation, and other production practices. It seems that some sort of classification scheme is needed to reduce the diversity of agriculture to a manageable set of variables that can be tied to water quality. Furthermore, changes in agricultural practices from year to year will have to be monitored, including chemical application rates. Will this information be obtained from aerial surveys, personal interviews with producers, or some other means? To ascertain agriculture's impacts on water quality, changes in other land-use activities and point source discharges must also be accounted for. Methods for obtaining this information on a periodic basis have to be outlined.

An important component of Cycle II that should support the understanding of agriculture's impacts on water quality trends is the greater emphasis on quantitative hydrologic analysis and the mathematical modeling of processes. The proposed HST is a national study team whose role will be the application and support of hydrologic and water quality models in NAWQA studies, including the use of models to aid in interpretive analysis of water quality. It seems logical that the HST will support research on the effects of changes in agricultural management on ground- and surface waters (Objectives *T5* and *T6*). In the Cycle II report (Gilliom et al., 2000), the USGS proposes that an Agricultural Chemical Source and Transport Team (ASTT) be formed to determine how physiographic characteristics and agricultural management practices affect runoff and recharge processes and the quality of shallow groundwater and streams. The Cycle II NIT report could be improved if it included a better presentation of how the HST and subsidiary ASTT will be coordinated with teams undertaking the various trend and understanding objectives. The committee provides the following related recommendations for Cycle II, based on recent changes in agriculture:

- Assess the impacts on water quality of state water quality protection pro-

grams (e.g., California's Pesticide Contamination Prevention Act, Nebraska's Ground Water Management and Protection Act, the Everglades Forever Act, North Carolina's Nutrient Sensitive Waters).
• Consider evaluating changes in sediment in areas where cropland has been retired by the Conservation Reserve Program.
• Examine the length of time between the introduction of a new pesticide and its appearance in water resources.
• Study changes in nitrogen and phosphorus loadings in regions where confined animal feeding operations have concentrated.
• Assess impacts on water quality of conservation tillage and other low-input agricultural practices.
• Consider tasking the proposed HST team to develop model(s) relating agriculture to water quality that can be applied in major agricultural areas.

## CONCLUSIONS AND RECOMMENDATIONS

The committee notes that many of the topics discussed in this chapter on the trend goal of NAWQA for Cycle II are pertinent to addressing the committee's statement of task (see Preface to this report). This is particularly true for the interrelated issues on extrapolation and aggregation of information at regional and national scales. In this regard, the USGS has long been a leader in the regionalization of hydrologic (especially streamflow) data, which has involved both aggregation and extrapolation efforts. As noted elsewhere in this report, NAWQA's consistent protocols and sampling regimen and the integrated approach to network design provide a good baseline that minimizes many problems that can make extrapolation and regionalization of its data and information problematic.

The USGS has a long and distinguished record of experience in hydrologic trends assessment. Indeed, many of the techniques currently employed for the estimation of trends in hydrologic time series are based on work by USGS scientists. The reliable and early detection of trends is of fundamental value because it can provide information on changes in water quality (especially related to anthropogenic sources) that might be useful for decision making and scientific understanding relating to the management of water quality. If trends are successfully detected in a timely fashion, along with a scientific understanding of the cause of those trends, it may be possible to implement management strategies to help reduce future degradation in water quality. Thus, information on a trend in water quality becomes particularly useful when the trend is linked to its underlying cause. To support this type of causal assessment, the Cycle II NAWQA program has established three themes and six objectives for determination of the trends in the status of water resources and the effects of urbanization and agricultural practices on water quality. The committee finds that the USGS and NAWQA are well positioned to carry out this important work in Cycle II. In this regard, NAQWA has established water quality baselines and monitoring networks in the Cycle I study units and is operating at time and spatial scales sufficient to establish these relationships. Several recommendations regarding the ability of the Cycle II NAWQA program to meet these trends and related objectives are provided below:

- NAWQA should continue its emphasis on an integrated-approach to water quality monitoring network design that attempts to coordinate efforts among various local, state, and federal agencies in an effort to make study unit designs as efficient and cost effective as is possible.
- The use of the NAUGLS approach introduced by Moss and Tasker (1991) should be extended to the design of water quality monitoring networks in Cycle II. This may be accomplished by extending the SPARROW multivariate statistical water quality model for use in the design of water quality monitoring networks similar to the way in which regional multivariate statistical models of various streamflow statistics have been used in the design of stream-gaging networks. NAWQA should develop a generalized quantitative approach to optimize the design of water quality monitoring networks that can be tailored to satisfy relevant NAWQA objectives.
- It is hoped that future NAWQA research will address all relevant issues relating to the objective of designing water quality monitoring networks to ensure accurate estimates of long-term loads. Guidance is needed on the trade-off that exists between cost of sampling and the precision associated with resulting estimates of long-term loads associated with a wide range of water quality constituents. Cost of sampling relates to the timing of these samples in both space and time and the availability of streamflow measurements. Accuracy of resulting long-term loads is dictated by the accuracy of the concentration-discharge relationship combined with the spatial and temporal frequency of water quality and quantity samples. Quantitative methods are needed to account for these issues. Extensive exploratory analyses are required to understand the probabilistic and mechanistic structure of water quality data collected by NAWQA.
- If trend evaluations are to be meaningful, they must account for all relevant statistical complexities to enable meaningful conclusions. Although most statistical complexities have already been dealt with, the issue of spatial correlation for regional inferences has not commonly been considered. Without a proper accounting for the spatial correlation of the water quality time series, resultant conclusions regarding trend assessments are likely be flawed. Future NAWQA research relating to detection of trends in water quality data should use methods similar to those described by Douglas et al. (2000) to account for the spatial correlation of water quality time series.
- NAWQA should place greater attention on the use of transfer function methods in its development of statistically based water quality models, because such methods properly account for the stochastic structure of time series of both water quality and its correlates.

- NAWQA should continue to emphasize practical and efficient models of the concentration-discharge relationship, because such models have applicability to trend assessments as well as many other important societal problems relating to water quality management. The committee also recommends that NAWQA place much greater emphasis in the future on the integration of physical and statistical models of the concentration-discharge relationship in the hopes that such integrative research will lead to much more credible and useful models.
- The reliance on observational data analysis for causal inferences is always subject to some controversy. Accordingly, NAWQA should consider application of recently developed methods in probability-based causal inference (e.g., Bayesian techniques) for its cause-and-effect studies.
- The NAWQA Cycle II study design has focused on current land-use conditions and their relationship to stream and groundwater attributes. As data analysis proceeds, attention must also be given to urban watershed management practices and land-use history so that land-use legacies can be incorporated.
- Habitat assessment and measures of fish and invertebrate communities appear to be more powerful measures of urban impact than proposed measures of the algal community, which are sensitive to recent storm events. Thus, if resources are limiting, the algal community analyses could be eliminated with the least loss to the program.
- The proposed NAWQA HST modeling team should include examination of strategies that combine information from site-specific data and regional model forecasts.
- To assess the water quality impacts related to agriculture, NAWQA should consider studies related to exemplary state water quality protection initiatives (e.g., California's Pesticide Contamination Prevention Act, Nebraska's Ground Water Management and Protection Act, the Everglades Forever Act, North Carolina's Nutrient Sensitive Waters).
- NAWQA should consider evaluating changes in sediment in areas where cropland has been retired by the Conservation Reserve Program.
- The length of time between the introduction of a new pesticide and its appearance in water resources should be examined.
- Changes in nitrogen and phosphorus loadings should be studied in regions where confined animal feeding operations have concentrated.
- The impacts on water quality of conservation tillage and other low-input agricultural practices should be assessed.

## REFERENCES

Barbash, J. E., G. P. Thelin, D. W. Kolpin, and R. J. Gilliom. 1999. Distribution of Major Herbicides in Ground Water of the United States. U.S. Geological Survey Water-Resources Investigations Report 98-4245. Sacramento, Calif.: U.S. Geological Survey.

Clarke, R. T. 1990. Statistical characteristics of some estimators of sediment and nutrient loadings. Water Resources Research 26(9):2229-2233.

Cohn, T. A. 1995. Recent advances in statistical methods for the estimation of sediment and nutrient transport in rivers. U.S. Natl. Rep. Int. Union Geod. Geophys. 1991-1994. Review of Geophysics 33:1117-1123.

Cohn, T. A., D. L. Caulder, E. J. Gilroy, L. D. Zynjuk, and R. M. Summers. 1992. The validity of a simple log-linear model for estimating fluvial constituent loads: An empirical study involving nutrient loads entering Chesapeake Bay. Water Resources Research 28(9):2353-2363.

Dixon, W., and B. Chiswell. 1996. Review of aquatic monitoring program design. Water Resources Research 30(9):1935-1948.

Douglas, E. M., R. M. Vogel, and C. N. Kroll. 2000. Trends in floods and low flows in the United States: Impact of spatial correlation. Journal of Hydrology 240:90-105.

Duffy, C. J., and J. Cusumano. 1998. A low-dimensional model for concentration-discharge dynamics in groundwater stream systems. Water Resources Research 34(9):2235-2247.

EPA (U.S. Environmental Protection Agency). 2000. The Quality of Our Nation's Water: A Summary of the National Water Quality Inventory. 1998 Report to Congress. EPA841-S-00-001. Washington, D.C.: U.S. Environmental Protection Agency, Office of Water.

Frick, E. A., D. J. Hippe, G. R. Buell, C. A. Couch, E. H. Hopkins, D. J. Wangsness, and J. W. Garrett. 1998. Water Quality in the Apalachicola-Chattahoochee-Flint River Basin, Georgia, Alabama, and Florida, 1992-95. U.S. Geological Survey Circular 1164. Reston, Va.: U.S. Geological Survey.

GAO (U.S. General Accounting Office). 1990. Greater EPA Leadership Needed to Reduce Nonpoint Source Pollution. GAO/RCED-91-10. Washington, D.C.: General Accounting Office.

GAO. 2000. Key EPA and State Decisions Limited by Inconsistent and Incomplete Data. GAO/RCED-00-54. Washington, D.C.: General Accounting Office.

Gilliom, R. J., K. Bencala, W. Bryant, C. A. Couch, N. M. Dubrovsky, L. Franke, D. Helsel, I. James, W. W. Lapham, D. Mueller, J. Stoner, M. A. Sylvester, W. G. Wilber, D. M. Wolock, and J. Zogorski. 2000. Study-Unit Design Guidelines for Cycle II of the National Water Quality Assessment (NAWQA). U.S. Geological Survey NAWQA Cycle II Implementation Team. Draft for internal review (11/22/2000). Sacramento, Calif.: U.S. Geological Survey.

Goolsby, D. A., W. A. Battaglin, G. B. Lawrence, R. S. Artz, B. T. Aulenbach, R. P. Hooper, D. R. Keeney, and G. J. Stensland. 1999. Flux and Sources of Nutrients in the Mississippi-Atchafalaya River Basin: Topic 3 Report. Silver Spring, Md.: National Oceanic and Atmospheric Administration.

Hamed, K. H., and A. R. Rao. 1998. A modified Mann-Kendall trend test for autocorrelated data. Journal of Hydrology 204:182-196.

Harding, J. S., E. F. Benfield, P. V. Bolstad, G. S. Helfman, and E. B. D. Jones III. 1998. Stream biodiversity: The ghost of land-use past. PNAS (US) 45:14843-14848.

Harmancioglu, N. B., and N. Alpaslan. 1992. Water quality monitoring network design: A problem of multi-objective decision making. Water Resources Bulletin 28(1):179-192.

Harmancioglu, N. B., O. Fistikoglu, S. D. Ozkul, V. P. Singh, and M. N. Alpaslan. 1999. Water Quality Monitoring Network Design, Water Science and Technology Library. Volume 33. Dordrecht, The Netherlands: Kluwer Academic Publishers.

Helsel, D. R., and R. M. Hirsch. 1992. Statistical Methods in Water Resources. Amsterdam: Elsevier.

Hirsch, R. M., and J. R. Slack. 1984. A nonparametric trend test for seasonal data with serial dependence. Water Resources Research 20(6):727-732.

Hirsch, R. M., W. M. Alley, and W. G. Wilber. 1988. Concepts for a National Water-Quality Assessment Program. U.S. Geological Survey Circular 1021. Reston, Va.: U.S. Geological Survey.

Hirsch, R. M., J. R. Slack, and R. A. Smith. 1982. Techniques of trend analysis for monthly water quality data. Water Resources Research 18:107-121.

Hooper, R. P., D. Goolsby, S. McKenzie, and D. Rickert. 1996. National Program Framework, U.S. Geological Survey, NASQAN Redesign. Reston, Va.: U.S. Geological Survey.

Lemke, K. A. 1991. Transfer-function model of suspended sediment concentrations. Water Resources Research 27(3):293-305.

McMahon, G., and T. F. Cuffney. 2000. Quantifying urban intensity in drainage basins for assessing stream ecological conditions. Journal of the American Water Resources Association 36:1247-1260.

Miller, C. R. 1951. Analysis of Flow-Duration, Sediment-Rating Curve Method of Computing Sediment Yield. Denver, Colo.: U.S. Bureau of Reclamation.

Moss, M. E., and G. D. Tasker. 1991. An intercomparison of hydrological network-design technologies. Hydrological Sciences Journal 36(3):209-221

MSEA (Management Systems Evaluation Area). 1995. Management Systems Evaluation Areas: Report of the Progress from 1990-1995. Washington, D.C.: U.S. Department of Agriculture.

Myers, D. N., and K. D. Metzker. 2000. Status and Trends in Suspended-Sediment Discharges, Soil Erosion, and Conservation Tillage in the Maumee River Basin—Ohio, Michigan, and Indiana. U.S. Geological Survey Water-Resources Investigations Report 00-4091. Columbus, Ohio: U.S. Geological Survey.

O'Connor, D. J. 1976. The concentration of dissolved solids and river flow. Water Resources Research 12(2):279-294.

Pearl, J. 2001. Causality: Models, Reasoning, and Inference. Cambridge, U.K.: Cambridge University Press.

Porter, S. D., T. F. Cuffney, M. E. Gurtz, and M. R. Meador. 1993. Methods for Collecting Algal Samples as Part of the National Water-Quality Assessment Program. U.S. Geological Survey Open-File Report 93-409. Raleigh, N.C.: U.S. Geological Survey.

Ribaudo, M. O., R. D. Horan, and M. E. Smith. 1999. Economics of Water Quality Protection from Nonpoint Sources: Theory and Practice. Agricultural Economic Report 782. Washington, D.C.: U.S. Department of Agriculture, Economic Research Service.

Sala, O. E., F. S. Chapin, J. J. Armesto, E. Berlow, J. Bloomfield, R. Dirzo, E. Huber-Sanwald, L. F. Huenneke, R. B. Jackson, A. Kinzig, R. Leemans, D. M. Lodge, H. A. Mooney, M. Oesterheld, N. L. Poff, M. T. Sykes, B. H. Walker, M. Walker, and D. H. Wall. 2000. Global biodiversity scenarios for the year 2100. Science 287:1770-1774.

Sanders, T. G., R. C. Ward, J. C. Loftis, T. D. Steele, D. D. Adrian, and V. Yevjevich. 1994. Design of Networks for Monitoring Water Quality. Littleton, Colo.: Water Resources Publications.

Smith, R. A., R. M. Hirsch, and J. R. Slack. 1982. A Study of Trends in Total Phosphorus Measurements at NASQAN Stations. U.S. Geological Survey Water Supply Paper 2190. Reston, Va.: U.S. Geological Survey.

Smith, R. A., R. B. Alexander, and M. G. Wolman. 1987. Water quality trends in the nation's rivers. Science 235:1607-1615.

Smith, R. A., R. B. Alexander, and K.J. Lanfear. 1993. Stream Water Quality in the Conterminous United States—Status and Trends of Selected Indicators During the 1980s, National Water Summary 1990-91. U.S. Geological Survey Water Supply Paper 2400. Reston, Va.: U.S. Geological Survey. Available online at http://water.usgs.gov/nwsum/sal/index.html.

Smith, R. A., G. E. Schwarz, and R. B. Alexander. 1997. Regional interpretation of water-quality monitoring data. Water Resources Research 33(12):2781-2798.

Smith, R. A. and R. B. Alexander. 1985. Trends in Concentrations of Dissolved Solids, Suspended Sediment, Total Phosphorus and Inorganic Nitrogen at U.S. Geological Survey National Stream Quality Accounting Network Stations. U.S. Geological Survey Water Supply Paper 2275. Reston, Va.: U.S. Geological Survey.

Spirtes, P., C. Glymour, and R. Scheines. 2001. Causation, Prediction, and Search. Cambridge, Mass.: MIT Press.

Swader, F. N. 1993. Agricultural research: The challenge in water quality. Pp. 16-20 in Proceedings of Agricultural Research to Protect Water Quality Conference, Minneapolis, February 21-24. Ankley, Iowa: Soil and Water Conservation Society.

USCB (U.S. Census Bureau). 2001. About Metropolitan Areas. Available online at http://www.census.gov/population/censusdata/urpop0090.txt.

Vecchia, S., G. Wiche, and W. Berkas. 1997. Evaluation of Sampling Procedures for Monitoring Trends in Surface Water Quality for the NAWQA Program. Internal draft. Bismark, N.D.: U.S. Geological Survey.

Wagner, B. 1997. Evaluation of Alternative Sampling Strategies for Long-Term Groundwater Quality Trend Detection in NAWQA. July 28. Internal draft. Menlo Park, Calif.: U.S. Geological Survey.

# 5

# NAWQA Cycle II Goals—Understanding

## INTRODUCTION

"Understanding" is the last of the three primary goals set for the National Water Quality Assessment (NAWQA) Program (with the status and trends goals described in Chapters 3 and 4, respectively). The NAWQA Cycle II Implementation Team (NIT) guidance document (Gilliom et al., 2000; see also Appendix A) states that this goal "is to improve explanation and understanding of: the sources of contaminants, their transport through the hydrologic system, the effects of contaminants and physical alterations on stream biota and ecosystems, . . . and implications for the quality of drinking water." To make progress in the realm of understanding the major factors that affect water quality, model development and application are essential. Understanding can be gained through the linkage of field studies with the analytical use of models, where observations are compared to a conceptual relationship expressed mathematically. Success of such a model in explaining observations is regarded as a measure of understanding the primary factors or mechanisms involved. Conversely, model development and application require understanding. The studies that provide an understanding of contaminant sources and their transport can be viewed as the raw materials for the design and development of water quality models. The previously described status and trends networks can provide such information, and therefore the proposed application of models in Cycle II rests firmly on the knowledge gained from Cycle I.

Models are developed to provide predictions of water quality conditions both spatially and temporally (i.e., through geographic extrapolation or prediction of future conditions). Detailed understanding of contaminant sources and their transport to water resources and an ability to predict future conditions are key to the development of efficient management and policies to protect the beneficial uses of the nation's water resources, including drinking water and viable ecosystems. As NAWQA progresses into Cycle II with an increased emphasis on its understanding goal, the importance of model application, as recommended by previous National Research Council (NRC, 1990, 1994) committees, should not be underestimated. Understanding and prediction, embodied in water quality models, are the cornerstones of water resources management for the future.

This chapter discusses four topics related to the goal of understanding cause and effect as related to water quality. First, the important role that models play in scientific understanding is explored. Understanding evolves from the thorough evaluation of data, and this is certainly one of the primary functions of model application. Models represent conceptual and mathematical relationships between the observations that we interpret as cause and effect. They are formulated at a wide range of temporal and spatial scales to represent a variety of phenomena—from rapid chemical reactions to long-term changes in the global environment. Each scale represents a different aggregation of controlling variables, so models are often not directly transferable between scales. This results in a hierarchy of models according to scale and highlights the importance of choosing models appropriately. They are also formulated in a variety of ways, including those based on mass-balance, statistical regression, and process-based (mechanistic) models. Regardless of the means of model formulation, there are several sources of uncertainty, including measurement uncertainty in the observations that serve as input and verification, model structure uncertainty, and parameter uncertainty. The quantification of uncertainty is important in providing perspective for interpretation of model results. Despite uncertainty, models play an important role in understanding causal factors that affect water quality, and model application is one way to illuminate the degree of understanding that exists. Different aspects of these topics are reviewed later in this chapter to highlight associated strengths and pitfalls in model application.

Second, the practical aspects of the overall proposed Cycle II implementation strategy are discussed. As a starting point, Cycle I of NAWQA naturally serves as the foundation for Cycle II. For example, key components of the surface water and groundwater monitoring assessments as well as the pilot-scale monitoring in lake and reservoir sediments conducted in Cycle I are carried over and expanded in Cycle II to form the National Trend Network for Streams, National Trend Network for Contaminants in Sediment, and National Trend Network for Ground Water (discussed in Chapter 4). Cycle I activities will also help form the planned Cycle II spatial studies of effects of land-use change on stream and groundwater quality. These trend assessments (both spatial and temporal) provide both the initial information needed to design the "targeted studies" proposed for Cycle II (see more below) and a context that gives a sense of their national priority (see in discussion of trends in Chapter 4). The two new components proposed for Cycle II that specifically reflect an increased emphasis on understanding and model application are:

- targeted water quality studies (that will variously address the six themes and 17 objectives of the understanding goal), and
- hydrologic systems analysis (HSA) to be guided by a newly formed Hydrologic Systems Team (HST).

As discussed in detail later, the plan is for "targeted studies" to focus on a limited set of the most important water quality topics and link these studies with other parts of the Cycle II design (Gilliom et al., 2000). Thus, each specific targeted study will be designed and executed by various topical teams composed of one or more planned Cycle II study units, and they will be assisted by a single, nationally responsible HST. These so-called targeted studies will focus on the major factors that govern water quality. The contaminants studied in Cycle I (e.g., pesticides, nutrients, volatile organic compounds) and the new drinking water source status assessments planned for Cycle II (see Chapter 3) are expected to provide the foundation for targeted study design wherever possible.

Third, the six themes and their corresponding objectives developed to describe the understanding goal (Gilliom et al., 2000; see also Appendix A) in Cycle II are assessed. These six themes were developed by categorization of the wide variety of scientific studies to be conducted by Cycle II study units to determine where commonalties or synergies between studies exist on a regional or national scale. Further perspective on the problem of setting goals and priorities on a national scale can be gained by relating the themes to the conceptual "source-transport-receptor" (STR) model described in the NIT guidance. The STR is a simple conceptual framework that can be used to organize a wide range of studies from across the nation to show both how they relate to each other and where information may be lacking. Although it is not explicitly stated in the latest NIT report, five of the six themes proposed for the understanding goal can be categorized according to the STR model as follows:

| Model Component | NIT Understanding Theme |
| --- | --- |
| Source | Sources of contaminants |
| Transport | Contaminant movement from land surface to groundwater |
| | Contaminant movement from land surface to surface water |
| | Groundwater-surface water interactions |
| Receptor | Effects on stream ecosystems |

The last understanding theme proposed for Cycle II, "extrapolation and forecasting," relates to the use of models for synthesis and prediction of water quality conditions for unsampled geographic areas (extrapolate) and future conditions (forecast). Indeed, it is the objective of Cycle II to "link the status and trend studies with an understanding of the natural and human factors that affect water quality" (Gilliom et al., 2000). Defining the relationships of the studies to each other suggests how regional- and national-scale understanding can be approached and is the first step in linking diverse information to use in setting future directions and in identifying gaps in understanding that may hinder extrapolation and forecasting. Thus, it is important that the targeted studies are interpreted in the context of these themes (or other conceptual frameworks) so that they may be evaluated in terms of geographic applicability and national priority. The discussion of extrapolation and forecasting and the many portions of this chapter that lead into it directly address the committee's statement-of-task issues (2) (extrapolation) and (4) (aggregation of information at regional and national scales) (see Chapters 1 and 8 for further information).

The fourth and final section includes the conclusions and recommendations for this chapter. The general conclusions of each section are followed by specific recommendations presented in the same order as the chapter sections. This chapter, focusing on the understanding goal for Cycle II and the related themes and objectives put forward by NAWQA, directly addresses the committee's statement-of-task issue (1), on methods for the improved understanding of causative factors affecting water quality conditions.

## ROLE OF MODELS IN UNDERSTANDING CAUSE AND EFFECT

### Introduction

Models may serve several functions with respect to understanding the causes and effects of water quality conditions. For example, they can be used as tools for diagnosis and explanation of the underlying mechanisms associated with the fate and transport of pollutants in the environment. They can serve as frameworks to integrate observations that vary in time and space and to predict the spatial and temporal distribution of future events. Models also are useful to focus and organize one's thinking about an environmental system. Good models should be consistent with current scientific understanding, and insofar as they are representative of reality, models provide a framework for organizing, communicating, and applying knowledge.

Model application, as recommended by the previous NRC committees, is gaining in importance as NAWQA progresses into Cycle II with an increased emphasis on the goal of understanding the major factors that affect observed water quality conditions (status) and trends. Understanding contaminant sources and contaminant transport to water resources is important for the development of effective management strategies to protect drinking water quality and the condition of stream ecosystems. Targeted studies and proposed model applications, which provide higher resolution (in time and space) than synoptic surveys, are important for identification and differentiation of natural and anthropogenic sources of contaminants in groundwater and surface water. Models are becoming increasingly important for water quality management, a notable example is the model used for the U.S. Environmental Protection Agency (EPA) Total Maximum Daily Load (TMDL) Program (NRC, 2001). For such applications, models may have a variety of uses that extend beyond their initial purpose(s).

Models exist in many forms. The models expected to play a major role in Cycle II of NAWQA are categorized into statistical models and analyses, geographic information system (GIS) analyses, process-based models, and hybrid statistical-GIS-process based models. Mathematical models are often characterized as either empirical or mechanistic (process oriented), even though most useful models have elements of both. In the purest sense, an empirical model is based on a statistical fit to data; the only substantive input is in the selection of predictor and response variables. Similarly, in pure form, a mechanistic model is a mathematical characterization of scientific understanding of the critical processes in the natural system; the only data input is in the selection of model constants, initial and boundary conditions.

## Mass Balance

The NAWQA Planning Team suggested that mass balance of constituents of concern could be utilized in a number of ways, especially related to explanatory science, a major activity proposed for Cycle II (Mallard et al., 1999). Simple mass-balance models will be utilized to help quantify the sources of contaminants to streams and aquifers. Mass balance (material balance) is frequently the beginning of quantitative assessment of the sources of nonpoint source (NPS) pollutants and their fate in the environment (e.g., Nolan, 1998; Vollenweider, 1975).

Figure 5-1 shows a simplistic approach to developing a mass balance. A control volume is defined, and rates of mass inputs and mass outputs cutting across the surface of the control volume are estimated. Next, transport within the control volume and across its borders are characterized, and sources and sinks of the constituent of concern within the control volume are defined. Lastly, the rate of change in mass per unit time in the control volume is obtained. For dissolved or particulate waterborne constituents, mass is defined as the product of water volume and constituent concentration. Thus, a mass balance of the water is also essential for waterborne constituents.

The spatial and temporal scales of a mass balance may vary over several orders of magnitude, depending on the objectives of the study in question. Control volumes may range from entire watersheds as in river basin models to infinitesimal slices as in numerical transport models. Other model characteristics also

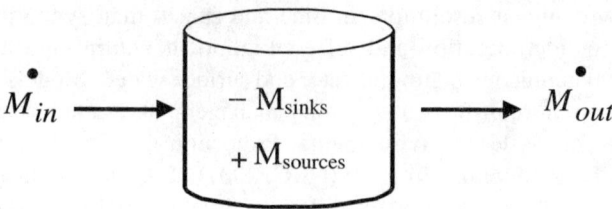

$$d(M)_{control}/dt = d(M)_{in}/dt - d(M)_{out}/dt + d(M)_{sources}/dt - d(M)_{sinks}/dt$$

FIGURE 5-1 Components of a simple mass-balance system and the continuity equation.

vary. When the entire watershed is not subdivided spatially, it is assumed to be spatially homogenous and the model is characterized as "lumped parameter." In contrast, when the watershed is subdivided, the model is characterized as "distributed parameter." Essentially, there are subcontrol volumes nested in a control volume for distributed parameter modeling. Distributed mass-balance models may be further subdivided into process-level mass-balance models. The pollutant may be considered conservative or reactive and assigned appropriate rates. Transport may be assumed to be convective (advective) or a combination of convective and diffusive (dispersive) flow, and the model may be deterministic or stochastic. The choice of constructing a particular mass balance is dictated in part by the study objectives and the availability of data. There is a continuum of mass-balance efforts ranging from conceptual mass balance of aggregated data to comprehensive numerical models that operate in small time and space intervals.

The requirements for constructing even a simplified, conceptual mass balance may entail substantial data and information collection. Take, for instance, determining sources of nitrogen in a stream where the land use is predominantly agricultural. The sources may include nitrogen inflow from the upstream reach, agricultural surface runoff, precipitation, discharge of collected subsurface drainage, groundwater discharge contributions into the stream, and so forth. The nitrogen species may include soluble (e.g., nitrate, ammonia, organic nitrogen) as well as particulate forms (e.g., sediment-bound nitrogen, plankton-bound nitrogen, other organic particulates). The nitrogen loading into the stream will have to be evaluated in terms of both concentration (mass per unit water volume) and mass (concentration × water volume). Thus, the nitrogen emission into a stream can become quite complex.

The examples taken from U.S. Geological Survey (USGS) studies in Box 5-1 illustrate the value of conducting mass balance for NPS pollutants. They not only provide significant insights into, and understanding of the problem situation, but also guide the design for further data collection, such as selection of fixed-site monitoring stations, and aid in the interpretation of the data collected. Moreover, the mass-balance approach may be used to track historical changes and provide predictions into the future with changes in management alternatives, land use, or environmental-ecological conditions, and so on.

There are also other simple conceptual frameworks that can be used to view mass balance data that give insight into the behavior of systems. These include the concepts of "retention" and "relative residence time." Retention is defined as the fraction of inflow mass that is retained in a control volume (such as a reservoir) and can be represented by the equation: $R$ = (inflow mass − outflow mass) ÷ inflow mass (Dillon and Rigler, 1974). This is an expression of net losses, or sedimentation of conservative (nonreactive) substances. Retention can be used to compare systems to each other and to determine its relationship with other factors. Residence time of a substance can be thought of as replacement rate and may be defined as $T$ = control mass ÷ inflow mass per unit time. For example, if a reservoir holds 100 million gallons (MG) of water and the inflow rate is 50 MG per year, then the residence time of water $T_w$ is (100 MG ÷ 50 MG per year), or 2 years. The residence time of a nutrient or pollutant can similarly be calculated. As discussed in Chapter 2, providing such simple measures for lakes and reservoirs where NAWQA data exist would be of benefit and likely stimulate further needed work.

In conclusion, conducting mass balances on constituents of concern is a worthy effort for Cycle II of NAWQA. Mass balances identify the sources and sinks of pollutants and their fate. They are a central tool for obtaining a better understanding of a problem situation. The committee recommends that at least a conceptual mass balance be developed for nonpoint source pollutants and constituents of concern for TMDLs in all Cycle II study units. The reactivity of a given NPS pollutant may be generic, but the extent of biological and chemical reactions of the pollutant is highly dependent upon site-specific conditions. Moreover, having such a knowledge base of mass-balance data is essential for evaluating the effectiveness of alternative best management practices to reduce pollutant loads. Depending on the availability of funding and expertise, more comprehensive mass-balance (e.g., hydrologic or hydraulic water quality) models might be applied to the more critical pollutants of selected study units.

## Statistical Models of Water Quality

As noted above, a statistical (or empirical) water quality model reflects greater attention to fitting the pattern in a data set than to describing mechanisms. Perhaps the simplest example of a statistical water quality model is the relation-

## BOX 5-1
### Examples to Illustrate the Utility of Models Used by USGS

The question of why mass balances and other model applications should be conducted is addressed below. Five examples of computations by NAWQA or other USGS study teams are briefly summarized to underscore the significance of their findings or implications.

- **Example 1.** Alexander et al. (2000) studied the effect of stream size on the delivery of nitrogen to the Gulf of Mexico. This study examined the flux of nitrogen from the Mississippi River to shallow waters of the Louisiana shelf in the Gulf of Mexico causing eutrophication and seasonal hypoxia. The average first-order loss rate of nitrogen shows rapid decline with channel size from 0.45 to 0.0005 per day. The proximity of nitrogen sources discharged into large streams and rivers is a major determinant of downstream transport of nitrogen to coastal waters.
- **Example 2.** Puckett et al. (1999) estimate nitrate contamination in an agro-ecosystem on an outwash aquifer using a nitrogen mass-balance budget. The Otter Tail outwash area (west central Minnesota) is intensively cropped. The nitrogen mass balance reveals that croplands contributed 89 percent of the excess total nitrogen, and fertilizers were the primary source of nitrogen leaching into the glacial outwash. If it were not for denitrification, the average nitrate leaching losses would be 19.3 instead of 10.8 kg of nitrogen per hectare-year.
- **Example 3.** McMahon and Woodside (1997) examined the nutrient mass balance for the Albemarle-Pamlico drainage basin in North Carolina and Virginia. The mass balance on total nitrogen and total phosphorus indicates the relative importance of agricultural NPS, the equal importance of atmospheric sources of nitrogen and phosphorus as compared to crop fertilizer, and the large amounts of nitrogen and phosphorus that are unaccountable or residual in the mass balance. The residuals reflect the uncertainty and error associated with estimating the various total nitrogen and total phosphorus mass-balance compartments.
- **Example 4.** Bachman and Phillips (1996) estimated the baseflow nitrogen load into the Chesapeake Bay. Nitrogen concentrations in well-drained soils were higher than in poorly drained soils. About 40 percent of the total nitrogen into the bay is from baseflow, indicating the importance of groundwater-stream interactions.
- **Example 5.** Smith et al. (1997) conducted a regional interpretation of a national water quality monitoring system. SPARROW was used to reduce problems of data interpretation caused by sparse sampling, network bias, and basin heterogeneity. The constructed regression models of total nitrogen and total phosphorus for monitoring 414 National Stream Quality Accounting Network streams throughout the conterminous United States provided insight into important sources and processes affecting nutrients in watersheds.

ship between concentration and streamflow, referred to as a rating curve model. Such models describe the often weak relationship between the concentration of a water quality constituent and river discharge. Rating curve models are used widely for extrapolating water quality loads when only a few measurements are available and for performing trend assessments. The widespread use of such statistical models is reviewed by Cohn (1995), who also discusses research developments and statistical issues involved in estimating loads from the concentration-discharge relationship. Since such rating curve models are used so widely, it is important that research be performed to improve these models. The data sets developed by NAWQA provide an excellent resource with which to test and improve rating curve-based water quality models.

As is often the case, improvements in one's ability to model the processes that give rise to water quality changes are likely to come from improvements in one's ability to understand those processes. Even a purely statistically based water quality model can be improved by introducing a physical basis. O'Connor (1976) used a deterministic approach to derive expressions for the spatial and temporal distribution of conservative dissolved solids in rivers in such a way that they could be expressed as rating curves. Similarly, Duffy and Cusumano (1998) document the physical situations that give rise to hysteretic (looped) behavior of concentration-discharge relationships. The USGS has historically made many contributions to the literature on statistical water quality (rating curve) models (see Cohn, 1995, for a review) with a focus on statistical innovations. Future improvements in rating curve and other statistically based water quality models are likely to come from the integration of statistical and physical process-oriented models. NAWQA should devote greater attention to research that seeks to improve the important concentration discharge model based on physical reasoning and physical process interpretations of the concentration discharge relationship of the type introduced by O'Connor (1976) and refined by Duffy and Cusumano (1998).

Statistical water quality models range from the very simple concentration-discharge models described above to the more complex regional multivariate statistical models such as SPARROW (Spatially Referenced Regression on Watershed Attributes). SPARROW is an effort to provide resource managers with spatially detailed information describing the location and magnitude of nutrient sources and watershed factors that affect the delivery of important chemical and biological constituents to receiving waters. Early applications of SPARROW to the Chesapeake Bay (Preston and Brakebill, 1999) and elsewhere have proven quite promising. The results of SPARROW modeling efforts led to an illustration of the spatial distribution of pollutant loads within the basin; such information is instrumental in watershed management programs that seek to target nutrient reduction areas. The committee recommends that NAWQA continue research relating to the improvement of SPARROW and to the application of this model to other watersheds in Cycle II. Continued comparisons should be performed between the output of SPARROW and analogous, yet more complex, process-oriented watershed-based water quality models. Also, as discussed in Chapter 4, SPARROW can be explored further for regionalization and extrapolation of NAWQA data.

## Process-Oriented Models of Water Quality

Process-based (or mechanistic) watershed models are mathematical representations of one's current understanding of hydrologic and water quality processes. A number of comprehensive, physically based watershed models now exist that are intended to simulate most watershed processes, ranging from hydrological state variables such as groundwater, streamflow, and evapotranspiration to water quality and sediment transport. A common feature of these models is the large number of parameters required relative to the available data. As a consequence, parameter selection becomes an art, and the models suffer from a general lack of rigorous testing. This suggests that the large process models are likely to be more useful for quantitatively representing processes and interactions in research studies than for forecasting the outcomes of proposed management actions. In addition, overparameterization means that the models lack an error term, so the research applications of these models is limited to informal analysis, because formal hypothesis testing is not truly possible.

One notable gap in process modeling identified in a recent NRC study (NRC, 2001) of the scientific basis for EPA's TMDL program is the lack of simple process models. Simulation models can, in principle, be crafted on a continuous scale of process detail, and there is nothing inherently correct, or better, about the scale of existing process models. The fact that current models are overparameterized suggests that simpler mechanistic expressions identifiable from the available data (in other words, those that support parameter estimation from the available data) should be considered as a viable alternative. One recent example of a model designed from this perspective is described in Borsuk et al. (2001); by avoiding overparameterization, Borsuk et al. were able to optimize the model fit to available data and also calculate a prediction error term. Accordingly, the committee recommends that process modeling within NAWQA avoid overparameterized models whenever possible.

To address problems of model selection, application, and analysis, a set of modular modeling tools, termed the Modular Modeling System (MMS; Leavesley, 1997) is being developed by the National Research Program of the USGS. The MMS approach attempts to enable a user to selectively couple the most appropriate process algorithms from applicable models to create an "optimal" model for the desired application. Where existing algorithms are not appropriate, new algorithms can be developed and easily added to the system. This modular approach to model development and application provides a flexible method for identifying the most appropriate modeling approaches given a specific set of user needs and constraints.

Nearly all of the programmatic objectives relating to the development of the MMS satisfy the NAWQA Cycle II objectives relating to the greater use of models for improving our understanding of the causes and effects of water quality. Integration of water quality-based modules within the MMS is a natural extension to both the MMS and the existing watershed-based water quality models. Furthermore, it may be quite possible to integrate water quality models that are not watershed based into the MMS modeling framework. The committee recommends that NAWQA attempt to merge developments in watershed based water quality models with the current Modular Modeling System, while striving to avoid overparameterized models.

### Uncertainty Must Be Considered

Water quality modeling is inherently uncertain. There is often significant uncertainty associated with the input, output, model structure, and parameters of water quality models. Uncertainty exists regardless of the model structure and regardless of whether the models have a statistical or process-oriented foundation. It is essential that all modeling studies account for the various sources of uncertainty; otherwise the results of such studies can be misunderstood. Statistical methods such as the simple rating curve approach and SPARROW can provide detailed uncertainty statements associated with estimates of long-term loads in the form of confidence intervals for predictions and/or standard errors associated with model parameters. As noted above, process-oriented modeling studies often lack sufficient data for parameter estimation; as a consequence, there is little basis for, and experience in, thorough error analysis with these models. The committee recommends that the HST attempt to quantify uncertainty associated with all aspects of its water quality modeling evaluations and that the HST develop a position paper on uncertainty.

## PROPOSED IMPLEMENTATION APPROACH FOR THE UNDERSTANDING GOAL OF CYCLE II

Several practical aspects of conducting scientific research have to be considered in an evaluation of the potential for successful implementation. In the case of NAWQA, these aspects include preliminary procedures such as defining objectives and selecting targeted studies to meet these objectives. Then there is a requirement for coordination of the hydrological and water quality models chosen for application. Sufficient staff, expertise, and financial support are all part of what is needed for successful implementation of a program, especially at the national scale. The degree of attention to each of these aspects will determine the success of Cycle II implementation.

The process of selecting topics for targeted studies to meet the understanding goal in Cycle II has not yet been clearly defined. The sequence of events suggested in the NIT planning document (Gilliom et al., 2000) begins with a statement of the 17 objectives that relate to the understanding goal. The implementation strategy to address these objectives is described in two main steps: (1) select topics for targeted studies to be addressed by topical study teams and (2) establish the HST to provide guidance in hydrologic and water quality model application. However, there is only a loose relationship between the 17 objectives and the subsequent four initial topic areas of targeted studies described in Gilliom et al. (2000). Although the topics suggested seem appropriate, they do not improve the focus since they relate to multiple objectives.

In the final section of the most recent NIT guidance document (Gilliom et al., 2000), an example design is presented for a targeted study topic on "sources and transport of agricultural chemicals into streams and shallow ground water" that will be designed and implemented by the Agricultural Chemical Source and Transport Team (ASTT). This planned targeted study includes a set of clearly defined hypotheses. The degree of focus demonstrated by this example is what seems to be lacking in the 17 objectives that are proposed. The example of the targeted study to be done by ASTT is one of the strongest parts of the planning document.

In relation to the many objectives, the committee suggests that more thought should be given to how information from multiple study units will be analyzed and integrated to address questions of national significance. The understanding objectives must be more clearly defined and stated in terms of testable hypotheses that can then be addressed by targeted studies designed for that purpose, as in the ASTT example. This probably means that objectives should not only be simplified, but also reduced in number.

At the May 2000 committee meeting at La Jolla, California, Dr. Robert Gilliom (USGS-NAWQA staff and lead of the NIT) provided a status report on the developing Cycle II themes, including the use of modeling tools. In that report, Gilliom stated that all Cycle II study units will utilize planning models, 75 percent of the study units will carry out mass-balance studies, 75 percent of the study units will conduct integrated field studies, 50 percent of the study units will conduct watershed flow and transport studies, and 25 percent of the study units will carry out comprehensive analysis and modeling of selected hydrologic cycle systems. A national Hydrologic Systems Team consisting of six to eight hands-on experts will be appointed to provide modeling guidance to the study units. It is clearly stated that the primary role of the HST is "not to provide technical support, but instead to provide leadership in conducting hydrologic systems studies" (Gilliom, 2000b).

The committee is concerned whether or not sufficient staff, resources, and expertise are allocated in the current planning to ensure the successful widespread use and application of models suggested for NAWQA. In the current plan, it is uncertain that the HST will be able to ensure the widespread use of models. Will each topical study team need its own subset of an HST to ensure that the model implementation is effective? Is there enough dedicated staff expertise within each study unit for this work? Model application is labor intensive, and to be successful, appropriate staff must be specifically designated for it.

Some of the committee's concern comes from a brief budget analysis presented in the NIT report (Gilliom et al., 2000), which suggests to the committee that the understanding goal is underfunded for the effort needed. However, it is not clear to the committee how various activities that crosscut the themes of Cycle II (i.e., support for modeling) are to be funded, nor is it the committee's task to deal in any detail with budget and resource issues. The NIT report recognizes this problem and clearly states that the understanding issues "can only partially be addressed by Cycle II investigations, even in all three groups of study units." With the expressed resource constraints, the number and complexity of the objectives for understanding and the staff resources and expertise issues must be carefully addressed in finalizing plans for targeted studies and modeling implementation.

The financial support available will ultimately govern what can be accomplished in Cycle II. As noted throughout this report, the USGS as an organization is uniquely qualified to pursue this national monitoring work. No other federal agency has the organization and infrastructure to collect, analyze, and report on the water resources of the nation with the long-term consistency critical to the success of this program. This capability has been repeatedly demonstrated through its Cycle I achievements (see also Chapters 1 and 8). Thus, the budget will be a key determinant of what can and will be accomplished in Cycle II.

## THEMES AND OBJECTIVES OF CYCLE II UNDERSTANDING GOAL

As noted previously, five of the six themes proposed for the understanding goal for Cycle II of NAWQA can be categorized according to the STR model. The sixth and last theme, "extrapolation and forecasting," relates to the use of models for the purposes of synthesis and prediction. The simple categorization of studies in this conceptual framework is a convenient way to make an initial assessment of where background information exists or where it is lacking. It provides a view of the status of current understanding, where commonalties or synergies between study units may exist that may benefit from coordination of modeling research, and what elements of understanding are in need of further investigation. The individual themes of the STR model are discussed below in terms of the proposed approaches and practical difficulties that should be considered in the design of studies to address the stated objectives. Ultimately, the information that arises from the various understanding themes (to be addressed by the formation of topical teams and targeted studies) will be assembled in the form of models.

## Sources of Contaminants

The four objectives that address the understanding theme "sources of contaminants" cover a variety of environmental scales and settings and are described in Gilliom et al. (2000):

*Objective U1:* Large-scale sources: Identify and quantify the most important large-scale natural and anthropogenic sources of selected contaminants to major streams and aquifers with mixed land-use influences and contaminant sources.

*Objective U2:* Urban and agricultural sources: Identify and quantify the most important sources of selected contaminants to recently recharged groundwater and to streams within urban and agricultural settings.

*Objective U3:* Spatial and temporal aspects of sources: Characterize and determine the relative importance of spatial, seasonal, and short- and long-term interannual variability of natural and anthropogenic sources of contaminants to surface and ground waters.

*Objective U4:* Mobilizing and metabolizing sources: Determine the relative influence of natural processes and human activities in creating contaminants or mobilizing naturally occurring contaminants.

Approaches to these four objectives are presented in an early version of the NIT report entitled *Preliminary Draft of Themes and Objectives for Cycle II Investigations—4/19/2000* (Gilliom, 2000b). The comments below refer to this version, despite its preliminary status, since the more recent Cycle II guidelines (Gilliom et al., 2000; see also Appendix A) do not provide any specific approaches for their implementation. Rather, the more recent guidance presents a broader strategy of forming the topical teams and the HST that will develop approaches (targeted studies) at some future time to address the understanding objectives. (Comments on the proposed teams as an approach are discussed in previous sections of this chapter.) Although this makes the current version of the guidance somewhat less specific, presumably some of the earlier ideas for implementing studies to address these four objectives will be considered by the topical teams. Therefore, some general comments on these are included below.

The proposed approach for Objective *U1* is that a large-scale mass-balance analysis will be developed for a selected basin (major stream) and/or major aquifer for selected constituents in most Cycle II study units (Gilliom, 2000b). The mass-balance analyses will be developed as fully as possible from Cycle I results and other existing data, with the objective of identifying and quantifying the primary sources of contaminants. In a subset of study units, the mass-balance analysis will be enhanced by additional data collection and model analysis.

As noted previously, the strength of the mass-balance approach is that it is based on conservation of mass and quickly highlights where observations are inadequate. However, the documentation of a balanced mass budget may not be simple depending on the resolution required. Several initial considerations for sampling design necessary to address this objective can be helpful. First, the timing of observations has important implications for how well they characterize a mass flow and therefore how complete a description they provide. For example, it is well known that for many substances, the majority of transport in a watershed takes place during a few major storm events each year; if the sampling program misses these critical times, an annual budget of inputs and outputs for a "control volume" will not be accurate. Another problem in producing a mass balance is that both water flow and mass concentration measurements are required; since concentration measurements are not routinely recorded as continuously as flow, interpolation methods will have to be applied to develop mass load estimates. A further complication is that many chemical and microbial contaminants, such as phosphorus or fecal coliforms, from different sources may be indistinguishable once they mix within a water mass. Sampling designs with appropriate spatial resolution may partially solve this problem, but in some cases, highly specific analyses may be necessary to identify different sources of the same contaminant; Boxes 5-2 and 5-3 provide examples of this and the use of simple mass balance for source identification.

There are some shortcomings of land-use data as a basis for making inferences about sources. Land use (as described in the environmental framework chapter [Chapter 1] of Gilliom et al., 2000; See Appendix A) can sometimes be used as a major explanatory factor for contaminant sources, but this does not account for sources of contaminants that are independent of land use (such as atmospheric inputs or wildlife sources of pathogens that are mobile). Routinely available land-use coverages (derived from remote sensing) may not provide high enough resolution to identify some sources, particularly anthropogenic point sources not apparent at the low resolution of land use. Another factor not accounted for by using an aggregated explanatory variable such as percentage land use is landscape position. The proximity of a contaminant source to a water resource may be very influential and more important than land use as an explanatory factor. High-resolution data at a subbasin scale and model application may be needed to obtain satisfactory understanding of large-scale natural and anthropogenic contaminant sources. The design for mass-balance estimation may also require indexing methods such that subbasin inputs can be estimated for those not measured because of the large geographic scales involved.

For Objective *U2*, the draft NIT guidance (Gilliom, 2000b) states that "for selected agricultural and urban areas, usually the most important nonpoint source areas identified in Objective [*U1*], sources within these land use areas will be specifically identified, and to the extent possible, quantified." For example, what are the sources of a particular pesticide to an urban stream, such as atmospheric deposition, lawn runoff, and spills? Although the approach is not clearly defined in the April 2000 NIT guidance, the Cycle I land-use surveys (LUSs) should provide insight into sources of contaminants from urban and agricultural land

> **BOX 5-2**
> **Sources of Coliform Bacteria;**
> **the Kensico Reservoir as a Case Study**
>
> With respect to the understanding theme of determining natural and anthropogenic sources of pollutants in water, the basic inventory of geology, land use, and land cover of the landscape provides an idea of potential sources of contamination. EPA's Source Water Assessment Program provides guidance for such evaluations. In addition, basic water quality monitoring provides information on the relative importance of different contaminants that actually reach the aquatic environment. In some cases, very specific supplementary analyses can be used to identify the sources of routinely monitored contaminants. For example, coliphage analyses can be used to supplement fecal coliform and *Escherichia coli* monitoring, and can give insight into whether the source is human, non-human, or some combination thereof. To take analytical specificity a step further, ribotyping (i.e., of ribosomal DNA from different sources of *E. coli*) can be used to identify specific animal sources of fecal contamination; however, substantial analytical effort is required to achieve this. An extensive reference library of ribotypes must be built (such as the one at the University of Pennsylvania) to interpret field data. Where highly specific "fingerprinting" analyses can be used, models may not be needed to distinguish between natural and anthropogenic contaminant sources.
>
> The source of fecal coliform contamination of Kensico Reservoir in the early 1990s was identified as an example of a "simple" screening-level analysis as defined by EPA (1997). Routine monitoring data showed

uses for recently recharged groundwater. The Trends Network and Land Use Studies will also provide a database on urban and agricultural sources, albeit at low resolution. In the statistical (regression) relation of GIS-derived land use to water quality concentration data, the sample data should be stratified between stormflow and baseflow and analyzed accordingly. Stormflow usually contains higher concentrations of many contaminants and often represents a greater proportion of surface runoff. The database should include a way to identify the type of flow that concentration measurements represent so that analyses are interpreted appropriately.

To understand the origin of contaminant sources in greater detail, supplementary high-resolution sampling may be needed. Sources of contaminants to streams from runoff and spills are best addressed by storm event sampling, rather than baseflow sampling. Sources of contaminants from atmospheric deposition

> several interesting features. A comparison of inflow and outflow data showed that fecal coliform concentrations leaving the reservoir were typically higher than those entering it; therefore the source of contamination was located within the local drainage basin. A second feature of the data was that peak concentrations occurred each autumn and were pronounced in the main basin of the reservoir. Field reconnaissance pointed to several possible sources including waterfowl, inflow tributaries draining suburban areas, septics, sewers, aqueduct inflows, and runoff from natural areas. Estimates of mass loadings from these sources suggested that quantitatively, they were about equally divided between waterfowl, storm water, and all other inputs. In addition, peak concentrations in the main basin of the reservoir coincided with the presence of migratory waterfowl that roosted there at night. As a remedial measure, a waterfowl harassment program was initiated in 1993 and coliform levels immediately dropped below Surface Water Treatment Rule limits for source waters. In this case, routine monitoring pointed to the location of coliform sources (i.e., originating within the basin), quantitative estimation of the mass loadings indicated the relative importance of sources, and correlation of waterfowl presence with elevated coliform concentrations pointed to a specific coliform source. Confirmation of the hypothesis that waterfowl were a major source of coliforms was done through coliphage sampling (i.e., a specific "fingerprinting" method). In this case, simple screening-level methods were sufficient to identify the problem and to suggest corrective management actions for an important drinking water reservoir of the New York City water supply system.

require direct monitoring of wet and dry deposition, and representative samples are not easy to obtain because of the spatial variability of weather and orographic effects. (Although atmospheric deposition as a source is mentioned as an approach for Objective $U2$, it seems to be more appropriate for $U1$, which will address large-scale phenomena.)

As noted previously, the example ASTT presents an excellent set of clearly articulated hypotheses for its proposed study (Gilliom et al., 2000). This example targeted study, designed to investigate sources and transport of agricultural chemicals, is an excellent model for the Cycle II topical teams to follow to develop focus for the objectives and further targeted study development.

The preliminarily proposed NIT approach (Gilliom, 2000b) for the characterization of variability ($U3$) is: "through existing data and Cycle I data analysis, describe the importance of historical changes in anthropogenic sources of con-

> **BOX 5-3**
> **Source Identification of Nutrients**
>
> An example of model application for the purpose of identifying nutrient sources is the phased approach to TMDL development for phosphorus that is currently in progress for New York City reservoirs. It began in 1994 with estimation of phosphorus loads from different sources in each basin using export coefficients, land use, and population density (as described in Reckhow et al., 1980). These loading estimates highlighted the relative importance of different sources in each of the reservoir drainage basins and therefore major sources that could be managed for nutrient reduction. The Vollenweider (1976) model (a steady-state model that relates loading to lake concentration) was then used to estimate the load needed to achieve a target lake concentration. Comparison of current and target loads demonstrated which basins needed nutrient reductions. More detailed hydrodynamic and eutrophication models are under development; these process-based models will allow calculation of phosphorus TMDLs to be defined in terms of more specific use-impairment measures than the general condition of eutrophication. It is hoped that these models will lead to phosphorus goals (TMDLs) based on distributions of chlorophyll or trihalomethane formation potential.

taminants to those currently present in surface and ground waters." Some contaminants that were applied or released at some time in the past may persist for long periods and only now be reaching receptors. Conversely, recent increases in contaminant use or release may not impact receptors for years or decades in the future. The proposed approach implies that Cycle I data should be analyzed using the space-for-time concept to predict future water quality and its distribution. This objective, however, should be stated separately from the idea of characterizing variability of contaminant sources. Variability of sources, both spatially and temporally, should be addressed first using existing data. Then the space-for-time concept can be explored, taking lag time (legacy) and exponential effects into account.

Although no approach was provided for Objective *U4* in either version of the NIT report, this objective strives to understand the role of humans in such water quality problems as acid rain, mercury accumulation in aquatic biota, polychlorinated biphenyl (PCB) deposition, and the impacts of all of these contaminants on the ecosystem. An important example of a situation related to this objective that NAWQA may wish to examine in Cycle II is the fate of PCBs in the Hudson

River. A recent EPA decision was made to dredge PCB-laden sediments from the river, and will present an opportunity to monitor conditions before and after remediation. Determining the relative influence of anthropogenic versus natural effects will require an understanding of these substances in terms of each component of the STR model. That is, models should be used as analytical tools to elucidate an understanding of the origin of contaminant sources, their transport characteristics, and their impacts on receptors. It will not be possible to manage these problems until such understanding is gained.

*Recommendations*

- NAWQA should consider evaluation of the variability of sources and contaminant occurrence as a first step for the objectives related to "understanding contaminant sources"; this should take place prior to subsequent space-for-time and targeted study design.
- Sampling design should consider interpolation and indexing considerations that may be necessary to develop as complete a mass balance as possible.
- For the identification of sources, when feasible, monitoring programs should employ highly specific or supplementary analyses that permit distinction between different sources of the same contaminant (e.g., coliphage or ribotyping of microbiological specimens, isotopic analyses of some chemicals).
- Some sampling strategies may require higher resolution to assess some contaminant sources. Sample data should be stratified between stormflow and baseflow to develop relationships between land use and stream concentrations for many contaminants.

## Contaminant Fate and Transport

There will be an increased emphasis in Cycle II on process-oriented studies and modeling to assess water quality at watershed scales. These models are directly applicable to the four objectives (*U5* to *U8*) that have been developed for better understanding of the transport processes by which contaminants move from the land surface to either surface water or groundwater (Gilliom et al., 2000). Research into the fate and transport of contaminants requires the use of tracer techniques (in the broad sense of the term) and ways to elucidate the age and flow rates of water masses where these cannot be readily measured. Since specific protocols for this were not presented in the latest NIT guidance, more general, conceptual comments are offered.

*Objective U5:* Saturated zone transport: Examine the extent to which the concentrations of specific contaminants in surficial aquifers are related to (a) types and distributions of land use in their recharge areas, (b) distributions of groundwater residence times, and (c) physical and biogeochemical processes in the saturated zone.

*Objective U6:* Unsaturated zone transport: Examine the influence of natural and anthropogenic factors on the concentrations and transformations of specific contaminants in the unsaturated zone, and on their flux through the unsaturated zone to shallow groundwater.

*Objective U7:* Land surface-to-stream transport: Determine how differences in watershed characteristics (e.g., soils, terrain, climate, geology) affect the transport of contaminants from watershed land surfaces into streams.

*Objective U8:* Within-stream transport: Determine how differences in characteristics of the stream within its catchment (e.g., terrain, climate) affect the transport of contaminants along streams.

The models and software packages to be used to help implement the above objectives by the planned HST in Cycle II will be coordinated with the needs of other NAWQA national synthesis teams as well as the study units. Regarding the HST, an October 2000 draft memorandum on hydrologic systems analysis and modeling planned for Cycle II (Gilliom, 2000a) states that the "primary focus of the HST will be on linking the most important hydrologic processes for each particular study area to a unified view of the nature and behavior of regional-scale hydrologic systems. Investigations will emphasize comparative analysis of the roles of common and unique aspects of hydrologic behavior in the wide range of hydrologic environments of the nation. Use of models will facilitate the development of comparative studies that are quantitative, objective, and hypothesis-driven." The committee supports this statement of the HST's focus. However, hypotheses cannot be tested without an error term, and this fact has implications (noted above) for the choice of models. Further, some uniformity in the use of models and software packages is requisite for national comparisons and aggregation of results. The HST should ensure such comparability.

In addition to these candidate models, Gilliom (2000a) states that simple mass-balance models will be constructed for major contaminants to help identify and quantify the sources of contaminants to streams and aquifers. Such mass-balance models may be simple, but as noted previously, they may pose difficulties in estimating sources and sinks, depending on available data and the resolution required for some contaminants.

*Ephemeral Streamflow*

Ephemeral streamflows can contribute contaminants to perennial streams and/or shallow aquifers. These contributions are more likely to be a significant issue, where these streams and aquifers are common. The current NAWQA program is not geared to the assessment of this type of contaminant transport, although some sampling of this nature has been conducted during Cycle I in the Central Arizona Basins Study Unit (William Wilber, USGS, personal communication, 2000; see Figure 1-1).

Contaminants can be contributed by two types of ephemeral streams: effluent dominated and flash flood dominated. The former includes streams where the flows are from wastewater treatment or reclamation plants (e.g., Las Vegas Wash in Las Vegas, Nevada). The latter includes flows (often flash floods) produced by precipitation; these flows may pose a more difficult sampling situation since they are irregular and often quite substantial. Ephemeral streams that are tributary to perennial streams can pose a significant threat to the perennial stream's water quality. Shallow groundwater systems may also be at risk since ephemeral flows are often a significant source of groundwater recharge in the arid West.

An example of perennial stream contamination by ephemeral flood flows was reported by Harwood (1995). This researcher studied the North Floodway Channel of Albuquerque, New Mexico, which drains an urban area of approximately 90 square miles. This ephemeral channel discharges to the Rio Grande, which is perennial in the Albuquerque area. Harwood found that the urban storm water contributions of dissolved zinc, lead, and aluminum were significant. Although the work was preliminary, the implication that other ephemeral channels in the Albuquerque area might contribute similar or greater amounts of contamination is clearly indicated. This scenario might be found in other parts of the arid West, where urban areas are drained by ephemeral streams that could be transporting a broad spectrum of contaminants (e.g., metals and organic compounds) to perennial streams or perhaps recharging them to shallow aquifers. In view of this gap in knowledge, NAWQA should develop strategies to sample ephemeral streamflows in Cycle II where possible, especially in perceived high-risk areas, such as those in which contamination could threaten perennial streams, lakes or reservoirs, or shallow groundwater systems.

*Recommendations*

- The HST should ensure some uniformity (and/or compatibility) in the use of models and software packages in Cycle II to enable national comparisons and aggregation of data and results. As previously discussed, to test research hypotheses, the HST must also address error terms and uncertainty in model selection and application.
- The current NAWQA program is not geared to the assessment of ephemeral streamflows that can be an important contaminant source in arid areas in particular. NAWQA should develop strategies to sample ephemeral streamflows, especially in perceived high-risk areas, such as those in which contamination could threaten perennial streams, lakes or reservoirs, or shallow groundwater systems.

## Groundwater-Surface Water Interactions

Our understanding of groundwater and surface water interactions has changed dramatically in recent years. Groundwater-surface water (GW-SW) interactions occur in all types of hydrogeologic and climatologic settings at all spatial and temporal scales (Winter et al., 1998). Groundwater can interact with streams, lakes, estuaries, bays, wetlands and coastal areas. The interconnection between groundwater and surface water means that they often behave as one reservoir and should be treated and managed as a single resource (Winter et al., 1998). Their interaction can have important ramifications for water quality because water, chemical constituents, and microorganisms can be transferred between the two reservoirs.

Although the NAWQA program has evolved from its original separate surface water and groundwater study units (Hirsch et al., 1988) there still needs to be more emphasis on GW-SW interactions and their effects on water quality in Cycle II. Indeed, one of the themes for Cycle II is GW-SW interactions with the associated question, What is the role of exchanges and interactions between groundwater and surface water in determining the degree and timing of contaminant levels and their effects on water quality? Associated with this theme are three objectives:

*Objective U9:* Large-scale groundwater-surface water interactions: Determine the influence of groundwater quality on stream quality and the influence of stream quality on groundwater quality at stream reach and larger scales in a broad range of environmental settings.

*Objective U10:* Small-scale groundwater-surface water interactions: Evaluate the effects of the riparian zone, including near-stream wetlands, various land uses or land covers and land management practices, on exchanges of water and associated chemicals in all directions between the land surface, shallow groundwater, hyporheic zone, and streams.

*Objective U11:* Hyporheic zone groundwater-surface water interactions: Increase understanding of the role of the hyporheic zone on the transport and fate of contaminants in both groundwater and surface water.

These objectives will be met in Cycle II through a variety of means (e.g., riparian zone mapping, baseflow synoptic studies, examination of stream stage changes), although groundwater flowpath studies occupy a prominent role in meeting each objective (Gilliom et al., 2000). During Cycle I, flowpath studies in many study units were dropped because of funding constraints and the lack of background information (Mallard et al., 1999). The flowpath studies conducted to date have emphasized relatively shallow systems with short flowpaths where groundwater discharges to streams. Recommendations by the NAWQA Cycle II

Planning Team (NPT) emphasize the importance of flowpath studies and encourage each Cycle II study unit to conduct at least one well-designed flowpath study.

The interface between groundwater and surface water systems is now recognized as a distinct zone, called the hyporheic zone (Gibert et al., 1990; Vervier et al., 1992). This refers to a region where active and dynamic exchange of water, microorganisms, and nutrients occurs between the surface water and the adjacent groundwater system (Gibert et al., 1990; Triska et al., 1989; Vervier et al., 1992). The hyporheic zone, or GW-SW ecotone, can be an important component in understanding surface water quality and near-surface groundwater quality in NAWQA study units (Hinkle et al., 2001).

Although there are study units that appear to be surface water dominated (e.g., Willamette Basin, New England Coastal Basins) and some that are groundwater dominated (e.g., Central Arizona Basins, High Plains Regional Ground Water Study), such outward appearances can be deceiving. Even in arid areas GW-SW interactions can be important. For example, one of the current thrusts of the USGS's Ground-Water Resources Program is the examination of GW-SW interactions in the arid Southwest. Similarly, a study unit ostensibly dominated by surface water can have important groundwater aspects—the Willamette Basin is an example (Hinkle et al., 2001). Groundwater-surface water interactions should be addressed in some Cycle II study units where they have not previously been considered (e.g., San Joaquin-Tulare River Basins).

Some excellent GW-SW interaction work has already emerged from Cycle I of NAWQA. For example, the nitrogen work in the Delmarva Peninsula by Bohlke and Denver (1995) showed that nitrate concentrations in two streams were strongly influenced by the nature of the adjacent groundwater flow systems (shallow versus deep circulation). Pesticide studies in the agricultural Eastern Iowa Basins (Squillace et al., 1993) indicated that GW-SW exchange has a profound influence on pesticide concentrations in both streams and the adjacent aquifers. McMahon and Bohlke (1996), working along the South Platte River Basin, showed that denitrification and GW-SW mixing within the alluvial aquifer sediments might greatly decrease the nitrate added to streams by discharging groundwater. Efforts such as these should be expanded, especially in agricultural areas.

Hyporheic zone investigations have also been conducted under NAWQA auspices. For example, Hinkle et al. (2001), working in the Willamette Basin Study Unit, showed that along a large (ninth-order) stream, significant nitrate uptake and reduction occurred along hyporheic flowpaths (see Box 5-4). They also noted strong vertical redox gradients and suggested that since nitrogen cycling is strongly influenced by redox conditions, nitrogen cycling is likely influenced by GW-SW interactions that control fluxes of redox species.

The emphasis on process-based models in Cycle II dictates the use of models that adequately represent GW-SW coupling, especially the hyporheic zone (GW-SW ecotone). This region of active interchange between the stream and groundwater can have steep gradients in a variety of important water quality

> **BOX 5-4**
> **Large-Scale Hyporheic Zone Studies**
>
> Hyporheic zone studies are frequently conducted at relatively small scales. However, in the Willamette Basin Study Unit, researchers have conducted a study that addresses hyporheic zone interactions on a large scale (Hinkle et al., 2001). This collaboration between the NAWQA Willamette Basin Study Unit group and the USGS National Research Program describes a multidisciplinary study (physical, isotopic, geochemical, and microbiological) of groundwater-surface water physicochemical interactions in a large (ninth-order) stream system. The integrated multidisciplinary approach is also significant in that it indicates that groundwater-surface water interaction water quality studies depend on a variety of techniques. The study suggests that hyporheic zone processes might be relevant in the management of the water quality of large streams. Large-stream hyporheic zones may possess the capability for nitrogen cycling (nitrate uptake and/or reduction; nitrification). Furthermore, nitrogen cycling in the hyporheic zone of large streams, especially near the water table, may be particularly sensitive to shifts in changing redox conditions due to dynamic groundwater-surface water interactions. Calculation of groundwater solute loads to large streams based on regional groundwater geochemistry, without accounting for hyporheic zone biogeochemical transformations, could result in significant errors. Research results may also have implications for water management issues other than just water quality. For example, uncertainties in streamflow measurements resulting from variations in proportions of river water in open channels versus hyporheic zones may have water rights or water use implications.

parameters (e.g., dissolved oxygen, redox conditions, dissolved organic and inorganic carbon, nitrogen species). It is imperative in Cycle II that NAWQA endeavors to include hyporheic processes in augmenting its understanding of water quality processes in GW-SW systems, as has been done in the Willamette Basin (Hinkle et al., 2001). Efforts to couple hyporheic zone modules to existing models should be accelerated, and the committee notes that a hyporheic module has recently been developed for MODFLOW. Since TOPMODEL (Beven, 1997) is often used in watershed studies, efforts might be directed toward adapting it to treat chemical transport. Some rudimentary attempts using TOPMODEL to simulate the transport of dissolved organic carbon (DOC) have already been made (Boyer et al., 1996; Hornberger et al., 1994) by coupling TOPMODEL to simple mixing-cell models. However, more work needs to be done. It may be appropriate to consider developing a hyporheic simulator for TOPMODEL. Where detailed data are lacking, simple hyporheic zone transport models can be used in data analysis (Bencala and Walters, 1983).

In conjunction with appropriate models, experiments should be designed to elucidate the water quality-controlling processes at the GW-SW interface (see Hinkle et al., 2001). Some of these experiments should be conducted in conjunction with the planned flowpath experiments for Cycle II. Tracer tests should be an integral part of these experiments, and the tests should be conducted in a variety of study units that subsume the land-use, hydrogeologic, and climatologic characteristics of the nation. Collaboration should be sought with individuals in USGS's Ground-Water Resources, Toxic Substances Hydrology, and National Research Programs and in the academic community (e.g., state Water Resources Research Institutes). Furthermore, study units should be encouraged to develop a process-based approach to the characterization of GW-SW interactions and their effects on water quality. If this means seeking out collaborators and cooperators, then that should be encouraged as well.

*Recommendation*

- The hyporheic zone, or groundwater-surface water ecotone, can be an important component in understanding surface water quality and near-surface groundwater quality in NAWQA study units. All appropriate Cycle II study units should endeavor to (1) design a process-based approach to characterize GW-SW interactions and their effects on water quality and (2) use process-based models that can include GW-SW interaction components to delineate the spatial and temporal variations in GW-SW interchange and the concomitant water quality changes. This may mean seeking out collaborators and cooperators to find the needed expertise and study sites, and this should be encouraged as well.

## Effects on Stream Ecosystems

As part of the targeted studies of factors governing water quality, three objectives relate directly to understanding the causes of degradation of stream ecosystems. Studies proposed under these objectives will assess the effects on stream biota and ecosystems of contaminants, contaminant mixtures, habitat modification, and other stressors (Gilliom et al., 2000).

*Objective U12:* Ecological effects of contaminants: Evaluate sediment or water toxicity at sites representing the range of environmental concentrations and mixtures of contaminants. At sites found to be toxic, determine concentrations and mixtures present in sediment and/or water and screen selected biota for physiological indicators of exposure such as biomarkers, as appropriate based on contaminants present.

*Objective U13:* Ecological effects of nutrient enrichment: Evaluate the relation between community structure and trophic dynamics among streams receiving varying levels of enrichment from allochthonous carbon, nitrogen, and phosphorus.

*Objective U14:* Ecological effects of habitat modification: Characterize and evaluate the relation between streamflow characteristics, physical habitat, and community structure.

The objective (*U12*) relating to understanding the ecological effects of contaminants is being considered in the first phase of Cycle II by assessing methyl mercury accumulation in aquatic organisms, a wise choice for a target contaminant (see Chapter 3 for detailed discussion of this pilot program). The committee thinks that this is an appropriate direction for NAWQA in Cycle II. Some of the other directions implied under the general description of Objective *U12* in Gilliom et al. (2000) sound like the beginnings of an ecotoxicology program, which would not be an appropriate course for NAWQA to take. It is not clear what studies would be done to evaluate "sediment or water toxicity at sites representing the range of environmental concentrations and mixtures of contaminants." To do this properly could consume most of NAWQA's resources in Cycle II. It would be inadvisable for NAWQA to embark upon a major ecotoxicology program since one already exists in USGS's Biological Resource Division (BRD). Thus, it is critical that Objective *U12* be clarified so that the types of studies being considered are clearly articulated and potential collaborative relationships with other internal and external (such as BRD) programs can be more clearly defined.

The effects of nutrient enrichment (*U13*) will be considered in the first phase of Cycle II, although these effects will be considered only in agricultural streams in this first phase (Gilliom et al., 2000). NAWQA justifies its focus on this objective by noting that the EPA will be developing regional nutrient criteria for streams during the time that the first monitoring phase of Cycle II is occurring. The committee feels that such an initial focus on Objective *U13* is a wise decision. The NAWQA data should be able to make a significant contribution to these important regulatory decisions that will have considerable impact on the nation's water quality in the future. However, focusing its attention only on nutrient enrichment in agricultural streams reduces its ability to contribute to the nationwide debate in all regions. Clearly there are nutrient enrichment problems in urban ecosystems as well, and regulations will not be able to ignore them. In addition, unless excess sedimentation is factored into this study design, its results will be of limited value (see Chapter 3 for further discussion). The presence of excess sediments not only alters turbidity (which is acknowledged in Gilliom et al., 2000), but also alters habitat available for algae. The response to excess nutrients will vary greatly depending on whether the stream has stable rock surfaces or unstable, shifting sediments. The presence of excess sediments in stream channels may be a consequence of present land use in the watershed or may be a legacy of past land use. In either case, sedimentation cannot be ignored. As discussed at length in Chapter 3, sediment is such an important water quality issue that it demands more attention from NAWQA and the USGS, both in direct status and trends measures (Chapter 4) and through other measures in ecological and habitat assessments.

Habitat degradation is often argued to be the most significant cause of ecological impairment (e.g., Naiman et al., 1995); thus it is appropriate that NAWQA have this as one of its objectives (*U14*). This represents an issue that will permeate most of the other attempts to synthesize ecological data in Cycle II. Thus, even though it will not be directly addressed through targeted studies in the first phase of Cycle II, it will be impossible to consider effects of nutrient enrichment (*U13*) or to consider effects of agricultural (e.g., Objective *T6*) or urban land-use change (e.g., Objective *T4*) without evaluating habitat degradation. The committee notes that considerable progress with this objective could and should be made using existing Cycle I data and the information being collected as part of the status and trends assessments.

Altered flow regime is often a significant component of habitat alterations. Because of its expertise in hydrology and hydrologic modeling, the USGS is particularly well suited to explore the linkages between flow regime, biological indicators, and ecosystem processes. Although this aspect of habitat alteration is included in the description of this objective, it does not appear that it will be investigated in the first group of Cycle II sites, except perhaps indirectly in studies associated with the urban gradient (Gilliom et al., 2000). Current approaches to understanding the relationships between altered flow regime, habitat alteration, and ecological impacts have relied on physical habitat simulation models (e.g., PHABSIM; Stalnaker et al., 1995) or on exploring essential components of the natural flow regime (Poff et al., 1997; Richter et al., 1996). USGS scientists have considerable data and expertise in this area and could make a significant contribution to understanding the linkages between flow alterations, aquatic biota, and water quality. Thus, NAWQA should consider including studies on the impact of alterations in hydrologic regime in its later-phase Cycle II plans.

The three objectives discussed here have evolved over the course of the committee's deliberations, and it is important to note that understanding the impacts of exotic species has been eliminated from the most recent list of Cycle II objectives. In many situations, it may be impossible to understand the factors resulting in degraded stream ecosystems without knowledge of the impacts of exotic species. There are also important feedbacks between degraded water quality conditions and vulnerability of stream assemblages to invasion by exotic species. The NAWQA data set could be used to make a valuable contribution to our understanding of this phenomenon. One important component of many biotic indices is some measure of the relative abundance of native versus exotic species, and this information should already be contained in the NAWQA database. Thus, although there may not be a specific objective relating to exotic species, data already collected and being analyzed by NAWQA will be valuable for examining these questions. NAWQA should find some way to encourage this synthesis, perhaps by developing cooperative arrangements with individuals working in the Fish and Wildlife Service's Invasive Species Program.

*Assessment of Biological Integrity*

Biological assessments focus on the living components of aquatic ecosystems and use those components to assess ecological health. They are typically used to help diagnose physical, chemical, and biological degradation of water resources and the cumulative effects of multiple sources of pollution. Because biological assessments are based on measures of the biota, which are continuously exposed to degraded water quality, they can provide a more integrative measure of water quality than simple grab samples taken for measures of water chemistry. Biologi-

---

**BOX 5-5**
**Biological Assessment Methods and Approaches**

Methods for biological assessment have made considerable progress since the first publication of the Index of Biotic Integrity (IBI; Karr, 1981). The first IBI was developed to assess fish in the Midwest, but it has since been modified to be used with fish in other parts of the country (Karr and Chu, 1999). A benthic IBI was developed for use with benthic macroinvertebrates based on data collected by the Tennessee Valley Authority (Kerans and Karr, 1994), and this metric has also been adapted for application in different parts of the United States. The IBI is a multimetric index developed from measured attributes of biological samples taken at sites representing a range of human influence. Attributes of the biological assemblage (e.g., taxa richness, relative number of sensitive taxa, relative abundance of different functional feeding groups) are scored, and those with the greatest ability to discriminate between sites are included in the metric. Usually 10-15 attributes of the assemblage are scored, and the scores are combined into a single IBI measure whose statistical properties are known (Karr and Chu, 1999).

A somewhat different approach to biological assessment is used in measures such as the North Carolina Biotic Index (Lenat, 1993). This approach is based on considerable knowledge of the sensitivity of different benthic macroinvertebrate taxa to pollutants, resulting from extensive collections in a wide range of environmental settings. Each taxon is given a sensitivity rating based on its tolerance to degraded conditions. An

cal assessments are an integral part of all themes of the NAWQA program—status, trends, and understanding.

Although extensive research has been conducted on methods and approaches for biological assessments (see Box 5-5), there is considerable debate not only about the best methods for assessing biological integrity ("biointegrity") of surface waters as required by the Clean Water Act, but also about the definition of biointegrity itself. In this regard, the USGS's NAWQA program is in excellent position to make a meaningful contribution to the methods debate based on results from Cycle I studies in which it has conducted bioassessments based on fish, benthic invertebrate, and algal assemblages. This should be a top priority of the recently formed Ecological Synthesis Team. The synthesis will be able to address questions of which indices and which taxonomic groups (fish, invertebrates, or algae) provide the most sensitive and robust assessment of biotic integrity based on a nationwide database that will be unique. This research has been much slower

---

index score for a site is given based on the tolerance values of the organisms collected at the site. A simple yet apparently widely applicable index based on sensitive organisms is the EPT richness index, which represents the total number of taxa found at a site in the following three predominantly benthic aquatic insect orders, chosen because of their sensitivity to pollutants: Ephemeroptera (mayflies), Plecoptera (stoneflies), and Trichoptera (caddisflies).

The Europeans and Australians have taken a somewhat different approach to biological assessment, which they have called RIVPACS (Wright, 1995). This method is based on extensive sampling of benthic macroinvertebrates in streams and rivers that are minimally impacted by humans. The biological assemblages, geographic locations, and physical attributes from these reference sites are combined in a multivariate statistical analysis that predicts the biological assemblage expected at a site based on its location. Samples collected from the site are then compared with this expected condition, and deviations from the expected can be attributed to anthropogenic effects. This approach is currently being explored in the western United States (Hawkins et al., 2000).

The use of algae in biological assessments has not been as widely applied as that of benthic macroinvertebrates or fish. Algae (more specifically, periphyton) respond fairly rapidly to changing conditions in streams and thus are less representative of longer-term conditions. Their abundance can be highly variable depending on time (e.g., since the most recent storm event). Nonetheless, biological indicators based on algal assemblages have been developed for different parts of the country.

than other aspects of NAWQA largely because of problems encountered with the processing and identification of the large number of samples collected in Cycle I, although these problems have now been resolved and the data backlog is being reduced. In particular, the Ecological Synthesis Team should determine if all of the biological assessments are necessary. For example, what insights to water quality conditions do invertebrate measures provide that are not provided by fish? What insights are provided by algal indices that are not apparent with other indices? This data set should be employed to suggest cost-effective ways to use biological monitoring in Cycle II when resource limitations do not permit everything to be measured.

In addition to evaluating different indices, it is critical that the Ecological Synthesis Team explore quantitative relationships between biotic indices and other measures of water quality. This topic is well suited for some rigorous modeling approaches. As state and local regulatory agencies incorporate more biological monitoring into their programs, it is crucial that the linkage between biological assessments and more traditional chemical measures of water quality be evaluated. Regulators need to know what leads to impairment of biological integrity and what must be regulated to reverse the trend of biological impairment. With its nationwide database, the USGS is in a position to make a meaningful contribution to this important issue.

One data set from a Cycle I study unit in which biological indicators were extensively studied (Cuffney et al., 1997) has revealed an interesting pattern of response that bears further investigation. Many of the indicators exhibited a threshold response to water quality degradation along a gradient of agricultural intensity; they declined precipitously at certain levels of an agricultural intensity index and then showed little change as agricultural intensity increased. Further investigations of the generality of this threshold response along other land-use gradients are planned as part of the Ecological Synthesis Team's work. This is an important direction for further research because, if thresholds are a general phenomenon, that could impact local, state, and national priorities for ecosystem rehabilitation. If a goal is to achieve maximum improvement in biological assemblages, these data would suggest that focusing rehabilitation efforts on sites near the threshold rather than on highly degraded sites would be most cost-effective.

*Recommendations*

- It would be inadvisable for NAWQA to embark upon a major ecotoxicology program that would overlap with the USGS's Biological Resources Division program. Thus, it is critical that Objective *U12* be clarified to better articulate proposed studies so that potential collaborative relationships with other programs can be more clearly defined.
- Focusing NAWQA's attention only on nutrient enrichment in agricultural

streams reduces its ability to contribute to the nationwide debate in all regions. Clearly, there are nutrient enrichment problems in urban ecosystems as well, and NAWQA should assess these as well, to the extent possible.

• Habitat degradation is often argued to be the most significant cause of ecological impairment. Thus, NAWQA should reconsider including studies on the impact of alterations in hydrologic regime and sediment transport in its later-phase Cycle II plans.

• Although there may not be a specific objective relating to exotic species, data already collected and being analyzed by NAWQA will be valuable for examining questions about their impacts. NAWQA should find ways to encourage this synthesis, perhaps by developing cooperative arrangements with the U.S. Fish and Wildlife Service's Invasive Species Program.

• The USGS's NAWQA program is in an excellent position to make a meaningful contribution to the debate on which biological indices are most meaningful based on results from Cycle I studies in which it has conducted bioassessments based on fish, benthic invertebrate, and algal assemblages. This should be a top priority of the Ecological Synthesis Team. Further, the information developed on which indices are most informative should be employed to suggest cost-effective ways to use biological monitoring in Cycle II. In addition to evaluating different indices, it is critical that the Ecological Synthesis Team explore quantitative relationships and potential threshold responses between biotic indices and other measures of water quality.

## Extrapolation and Forecasting

Geographic extrapolation of analyses from NAWQA study units to unmeasured areas and forecasting future water conditions (e.g., based on land management changes) is an essential feature of NAWQA. In the Cycle I national synthesis reports (see also Chapter 6), scientific inferences tend to be expressed in terms of "percent of the samples," which in effect restricts the geographic coverage to NAWQA watersheds. Strictly speaking, this is incompatible with the initial NAWQA program goal "to describe the status and trends in the quality of the nation's ground- and surface-water resources," but it is nonetheless a prudent strategy given the limited basis for extrapolation to unsampled areas (representativeness issues that are fundamental to extrapolation considerations are also discussed in Chapters 2 and 4).

It is noteworthy that the Cycle II goals have been scaled back to describing "water-quality conditions for a *large part* of the nation's water resources" (emphasis added), while three Cycle II understanding objectives (*U15* to *U17*; see below) explicitly address extrapolation and forecasting. These are both improvements to NAWQA; the first acknowledges the limitations in coverage of the NAWQA study units, and the second targets a scientific basis for inferences beyond the study unit site samples.

*Objective U15:* Evaluation of empirical extrapolation and forecasting models: Develop, evaluate, and improve empirical models for spatial extrapolation and forecasting using statistically based methods such as regression analysis.

*Objective U16:* Evaluation of deterministic extrapolation and forecasting models: Systematically evaluate and test selected existing simulation models for their potential value in extrapolation and forecasting of water quality in streams and groundwater.

*Objective U17:* Application of extrapolation and forecasting models: Apply the most appropriate empirical and deterministic models to specific extrapolation and forecasting objectives.

The three understanding objectives for Cycle II address extrapolation to unmeasured geographic areas and future forecasting. Based on the brief statement and description of objectives in the NIT guidance (Gilliom et al., 2000; see Appendix A), empirical and mechanistic (simulation) models are expected to play a key role in this extrapolation and forecasting theme. Beyond that, no additional details are presented in the Cycle II implementation guidelines to describe how this will be accomplished, except that consideration will be given to "knowledge of land use and contaminant sources, natural characteristics of the land and hydrologic system, and our understanding of governing processes."

Despite the absence of details on extrapolation and forecasting planned for Cycle II, several papers and reports have been prepared by USGS-NAWQA scientists that provide an indication of related techniques that may be applied in Cycle II. For example, Nolan et al. (1997) mapped the United States based on nitrate input risk groups to statistically assess (using boxplots and analysis-of-variance techniques) the relationship between risk group and groundwater nitrate concentration (based on NAWQA measurements). Black et al. (2000), Rupert (1998), and Tesoriero and Voss (1997) all used logistic regression to develop statistical models to predict the probability of contaminant (nitrate and pesticides) concentration exceedances in surface and groundwater from land use and various land features.

These articles indicate a consistent strategy to develop statistical relationships between NAWQA water quality data and predictor variables that are widely measured and reported. This is a prudent strategy. However, NAWQA scientists have recognized the importance of technical support for modeling and hydrologic system analysis with plans to establish the HST (Gilliom et al., 2000). In a similar manner, NAWQA should establish a statistical model support team (perhaps as a subgroup within HST) to provide guidance on the selection and application of these increasingly important modeling tools. This should be an easy task for USGS to support, since some of the best hydrologic statisticians in the United States already are employed by the USGS. As noted in Chapter 4, USGS has long been a leader in "regionalization" and extrapolation, of streamflow data in particular. The statistical model support group should build on this experience to adapt methods for water quality assessments. Among the issues that this proposed group could advise on are (1) conditions permitting extrapolation of statistical model inferences to unsampled areas, (2) alternative regression and multivariate methods such as smoothing and robust estimation, (3) consequences of extensive model specification searching, and (4) statistical support for causal conclusions and application of emerging methods in causality assessment (e.g., Spirtes et al., 2001). These specifications parallel the prior recommendations for a focus on parsimonious and adaptive models; including Bayesian and Kalman filtering analysis.

*Recommendation*

- NAWQA should establish a statistical model support team (perhaps as a subgroup within the HST) to provide guidance on the selection and application of the increasingly important modeling tools. This group can build on the USGS's historical strength and experience in regionalization and extrapolation of streamflow. Also the group should explore and advise on various issues, including more parsimonious and adaptive models, new techniques (e.g., Bayesian), and previously discussed issues such as uncertainty assessment.

## CONCLUSIONS AND RECOMMENDATIONS

The application of water quality models is an essential activity to achieve the understanding goal of Cycle II NAWQA. The status and trends networks provide the basis for modeling contaminant sources and their transport, and therefore, the proposed application of models in Cycle II will rely heavily on the knowledge gained from Cycle I. Modeling of contaminant sources and their transport to water resources, and hence an ability to predict future conditions, are key to the development of efficient management and policies to protect the beneficial uses of the nation's water resources, including drinking water and viable ecosystems. Therefore, the application of water quality models is the cornerstone of water resource management for the future, and this was clearly recognized by the NRC as early as 1990 when model application was recommended for NAWQA.

Several types of models are proposed for application in Cycle II. The committee presents an overview of models, including those categorized as conceptual, mass balance, statistical regression, process based (mechanistic), and hybrids of these. The systematic arrangement of these models in a hierarchy of spatial and temporal scales can be linked to, and will advance the development of, the Modular Modeling System. When more complex models are applied, the focus should be on those that can be parameterized with data. Once models are applied, the quantification of uncertainty is important to provide perspective for interpretation of model results; therefore, written guidance to standardize the documentation of uncertainty should be developed.

Successful implementation of model application will require detailed planning to accommodate the practical aspects of the Cycle II implementation strategy. The practical aspects to be dealt with include a clear statement of objectives, coordination of hydrological and water quality models, designation of appropriate staff resources, and optimization of financial support.

The six themes with 17 objectives that describe the understanding goal for Cycle II were developed (in the preliminary planning) by categorization of the wide variety of scientific studies proposed by the Cycle II study units. This is the first step needed to determine where commonalties or synergies exist between studies on a regional or national scale. From this, topics are to be selected for targeted studies to meet the understanding goal, but the selection process has not yet been clearly defined. A greater distillation of this collection of ideas is needed since it does not appear that the objectives have been sufficiently refined and focused. The final section of the NIT guidance (Gilliom et al., 2000) presents an example for targeted study development by an Agricultural Chemical Source and Transport Team. This is a strong example that included a set of more clearly defined hypotheses for study. NAWQA should build on this example to further define and clarify study development. It also may be useful to relate the proposed studies and their components to the conceptual source-transport-receptor model presented in the NIT guidance. The STR could be used to organize a wide range of studies from across the nation to show how they relate to each other and where information may be lacking. It is important that the targeted studies are interpreted in the context of NAWQA's themes and related objectives for Cycle II so they may be evaluated in terms of geographic applicability and national priority.

At this important juncture in the development of NAWQA, the committee concludes that the USGS has several major opportunities to advance scientific understanding of factors that affect water quality conditions. As noted throughout this report, the USGS as an organization is uniquely qualified to pursue this national monitoring work. No other federal agency has the organization and infrastructure to collect, analyze, and report on the water resources of the nation with the long-term consistency critical to the success of this program. However, the committee is concerned whether sufficient staff, resources, and expertise are allocated to ensure that modeling implementation can be accomplished and that targeted studies can be adequately developed. Although the NIT report (Gilliom et al., 2000) recognizes the resource problem, its resolution will ultimately govern what can be accomplished in Cycle II.

Given these general conclusions from the review of NAWQA plans for the understanding goal in Cycle II, there are many specific recommendations that the committee offers; these are listed below in order of the chapter sections.

## Recommendations for Modeling Approach

- Conducting mass balances on constituents of concern is a worthy effort in Cycle II. It is strongly recommended that at least a conceptual mass balance be developed for nonpoint source pollutants studied in each study unit.
- The committee recommends, as discussed in Chapters 2 and 3, that although lakes and reservoirs are not a prominent focus of NAWQA, there are opportunities to use a mass-balance approach to examine and compare their behavior. NAWQA should use the data collected to summarize retention and residence time for key contaminants. This will contribute to understanding system behavior and provide data to stimulate further studies of controlling factors of water quality.
- The USGS and NAWQA should continue research on improving rating-curve and other statistically based water quality models. Future improvements are likely to come from the integration of statistical and physical process-oriented models. Research should focus on physical reasoning and physical process interpretations of the concentration-discharge relationship of the type introduced by O'Connor (1976) and Duffy and Cusumano (1998).
- Early applications of SPARROW (e.g., to the Chesapeake Bay; Preston and Brakebill, 1999) have proven quite promising. The committee encourages NAWQA to continue research relating to the improvement and application of SPARROW.
- NAWQA-USGS should focus on simple, parsimonious process models (i.e., models that are not overparameterized) where parameter estimation and mechanistic expressions can relate to available data. Further, consistent with the recent NRC study of the scientific basis of EPA's TMDL program (NRC, 2001), NAWQA should encourage that such models be usable in an adaptive framework. This includes techniques such as Bayesian analysis, Kalman filtering, and data assimilation that allow model forecasts to be updated and improved over time with monitoring data.
- All future NAWQA studies should attempt to evaluate the uncertainty associated with all aspects of their water quality modeling evaluations, and the Hydrologic Systems Team should develop a guidance document on this topic.

## Recommendations for Implementation

- The 17 objectives for the understanding goal, as presented in the NIT report, are complex and address too many issues to have a good focus. Further, the process of selecting topics for targeted studies to meet the understanding goal in Cycle II has not yet been clearly defined. The example of the ASTT, however, was well done, presenting clearly defined hypotheses and a focus lacking in other areas. The committee recommends that NAWQA build on this example to develop and refine its targeted study approach.

- NAWQA must resolve staffing, expertise, and resource issues if the objectives of the understanding goal are to be successfully addressed in Cycle II. It is unclear what portion of the objectives can be met at the level of resources suggested in planning documents and whether sufficient staff expertise is available for model application and implementation.

## Recommendations on Understanding Themes and Objectives

*Sources of Contaminants*

- NAWQA should consider evaluation of the variability of sources and contaminant occurrence as a first step for the objectives related to "understanding contaminant sources"; this should take place prior to subsequent space-for-time and targeted study design. NAWQA should also consider collaboration with other agencies that may have pertinent data to evaluate statistical properties of contaminant variability.
- Sampling design should include interpolation and indexing considerations that may be necessary to develop as complete a mass balance as possible.
- For the identification of sources, when feasible, monitoring programs should employ highly specific or supplementary analyses that permit distinction between different sources of the same contaminant (e.g., coliphage or ribotyping of microbiological specimens, isotopic analyses of some chemicals).
- Some sampling strategies may require higher resolution to assess some contaminant sources. Sample data should be stratified between stormflow and baseflow for the development of relationships between land use and stream concentrations for many contaminants.

*Contaminant Fate and Transport*

- The HST should ensure some uniformity (and/or compatibility) in the use of models and software packages in Cycle II to enable national comparisons and aggregation of data and results. As previously discussed, to test research hypotheses the HST must also address error terms and uncertainty in model selection and application.
- The current NAWQA program is not geared to the assessment of ephemeral streamflows that can be an important contaminant source in arid areas in particular. NAWQA should develop strategies to sample ephemeral streamflows, especially in perceived high-risk areas, such as those in which contamination could threaten perennial streams, lakes or reservoirs, or shallow groundwater systems.

## Groundwater-Surface Water Interactions

• The hyporheic zone, or groundwater-surface water ecotone, is an important component in understanding surface water quality and near-surface groundwater quality. All appropriate Cycle II study units should endeavor to (1) design a process-based approach to characterize GW-SW interactions and their effects on water quality and (2) use process-based models that can include GW-SW interaction components to delineate the spatial and temporal variations in GW-SW interchange and the concomitant water quality changes. This may mean seeking out collaborators and cooperators to find the needed expertise and study sites, and this should be encouraged as well.

## Effects on Stream Ecosystems and Assessments of Biological Integrity

• It would be inadvisable for NAWQA to embark upon a major ecotoxicology program that would overlap with the USGS's Biological Resources Division program. Thus, it is critical that Objective *U12* be clarified to better articulate proposed studies so that potential collaborative relationships with other programs can be more clearly defined.

• Focusing NAWQA's attention only on nutrient enrichment in agricultural streams reduces its ability to contribute to the nationwide debate in all regions. Clearly, there are nutrient enrichment problems in urban ecosystems as well, and NAWQA should assess these to the extent possible.

• Habitat degradation is often argued to be the most significant cause of ecological impairment. Thus, NAWQA should reconsider including studies on the impact of alterations in hydrologic regime and sediment transport in its later-phase Cycle II plans.

• Although there may not be a specific objective relating to exotic species, data already collected and being analyzed by NAWQA will be valuable for examining questions about their impacts. NAWQA should find ways to encourage this synthesis, perhaps by developing cooperative arrangements with the U.S. Fish and Wildlife Service's Invasive Species Program, for example.

• The USGS's NAWQA program is in excellent position to make a meaningful contribution to the debate on which biological indices are most meaningful based on results from Cycle I studies in which it has conducted bioassessments based on fish, benthic invertebrate, and algal assemblages. This should be a top priority of the Ecological Synthesis Team. Further, the information developed on which indices are most informative should be employed to suggest cost-effective ways to use biological monitoring in Cycle II. In addition to evaluating different indices, it is critical that the Ecological Synthesis Team explore quantitative relationships and potential threshold responses between biotic indices and other measures of water quality.

*Extrapolation and Forecasting*

- NAWQA should establish a statistical model support team (perhaps as a subgroup within the HST) to provide guidance on the selection and application of the increasingly important modeling tools. This group can build on the USGS's historical strength and experience in regionalization and extrapolation of streamflow. Also the group should explore and advise on various issues, including more parsimonious and adaptive models, new techniques (e.g., Bayesian), and previously discussed issues such as uncertainty assessment.

## REFERENCES

Alexander, R. B., R. A. Smith, and G. E. Schwarz. 2000. Effects of stream channel size on the delivery of nitrogen to the Gulf of Mexico. Nature 403:758-761.

Bachman, L. J., and P. J. Phillips. 1996. Hydrologic landscapes on the Delmarva Peninsula, Part 2, Estimates of base-flow nitrogen load to Chesapeake Bay. Journal of American Water Resources Association 32(4):779-791.

Bencala, K. E., and R. A. Walters. 1983. Simulation of solute transport in mountain pool-and-riffle stream: A transient storage model. Water Resources Research 19:718-724.

Beven, K. J. 1997. Distributed Hydrological Modelling: Applications of the TOPMODEL Concept. New York: Wiley & Sons.

Black, R. W., A. L. Haggland, and F. D. Voss. 2000. Predicting the probability of detecting organochlorine pesticides and polychlorinated biphenyls in stream systems on the basis of land use in the Pacific Northwest, USA. Environmental Toxicology and Chemistry 19:1044-1054.

Bohlke, J. K., and J. M. Denver. 1995. Combined use of groundwater dating, chemical and isotopic analyses to resolve the history and fate of nitrate contamination in two agricultural watersheds, Atlantic coastal plain, Maryland. Water Resources Research 31(9):2319-2339.

Borsuk, M. E., D. Higdon, C. A. Stow, and K. H. Reckhow. 2001. A Bayesian hierarchical model to predict benthic oxygen demand from organic matter loading in estuaries and coastal zones. Ecological Modelling 143:165-181.

Boyer, E. W., G. M. Hornberger, K. E. Bencala, and D. M. McKnight. 1996. Overview of a simple model describing variation of dissolved organic carbon in an upland catchment. Ecological Modelling 86:183-188.

Cohn, T. A. 1995. Recent advances in statistical methods for the estimation of sediment and nutrient transport in rivers. In Vogel, R. M. (ed.) U.S. National Report 1991-1994—Contributions in Hydrology. Washington, D.C.: American Geophysical Union. Available online at http://www.agu.org/revgeophys/cohn01/cohn01.html.

Cuffney, T. F., M. R. Meador, S. D. Porter, and M. E. Gurtz. 1997. Distribution of fish, benthic invertebrate and algal communities in relation to physical and chemical conditions, Yakima River basin, Washington, 1990. U.S. Geological Survey Water-Resources Investigations Report 96-4280. Raleigh, N.C.: U.S. Geological Survey.

Dillon, P. J., and F. H. Rigler. 1974. A test of a simple nutrient budget model predicting the phosphorus concentration in lake water. Journal of the Fisheries Research Board of Canada 31:1771-1778.

Duffy, C. J., and J. Cusumano. 1998. A low-dimensional model for concentration-discharge dynamics in groundwater stream systems. Water Resources Research 34(9):2235-2247.

EPA (U. S. Environmental Protection Agency). 1997. New York City Watershed: Filtration Avoidance Determination. Mid-Course Review. Available online at http://www.epa.gov/r02earth/water/nycshed/fadrev2.htm.

Gibert, J., M. J. Dole-Olivier, P. Marmonier, and P. Vervier. 1990. Surface water/groundwater ecotones. Pp. 199-225 in Naiman, R. J., and H. Decamps (eds.) Ecology and Management of Aquatic-Terrestrial Ecotones. Park Ridge, N.J.: The Parthenon Publishing Group.

Gilliom, R. J. 2000a. Hydrologic systems analysis and modeling in Cycle II of NAWQA. Draft memo, submitted to the NRC committee, October 3.

Gilliom, R. J. 2000b. Preliminary draft of themes and objectives for Cycle II investigations. Draft memo, submitted to NRC committee, April 19.

Gilliom, R. J., K. Bencala, W. Bryant, C. A. Couch, N. M. Dubrovsky, L. Franke, D. Helsel, I. James, W. W. Lapham, D. Mueller, J. Stoner, M. A. Sylvester, W. G. Wilber, D. M. Wolock, and J. Zogorski. 2000. Study-Unit Design Guidelines for Cycle II of the National Water Quality Assessment (NAWQA). U.S. Geological Survey NAWQA Cycle II Implementation Team. Draft for internal review (11/22/2000). Sacramento, Calif.: U.S. Geological Survey.

Harwood, A. K. 1995. The Urban Stormwater Contribution of Dissolved Trace Metals from the North Floodway Channel, Albuquerque, NM, to the Rio Grande. Professional project report, Master of Water Resources Administration Degree, University of New Mexico.

Hawkins, C. P., R. H. Norris, J. N. Hogue, and J. W. Feminella. 2000. Development and evaluation of predictive models for measuring the biological integrity of streams. Ecological Applications 10:1456-1477.

Hinkle, S., J. Duff, F. Triska, A. Laenen, E. Gates, K. Bencala, D. Wentz, and S. Silva. 2001. Linking hyporheic flow and nitrogen cycling near a large river in the Willamette Basin, Oregon. Journal of Hydrology 244(3-4):157-180.

Hirsch, R. M., W. M. Alley, and W. G. Wilber. 1988. Concepts for a National Water-Quality Assessment Program. U.S. Geological Survey Circular 1021. Denver, Colo.: U.S. Geological Survey.

Hornberger, G. M., K. E. Bencala and D. M. McKnight. 1994. Hydrological controls on dissolved organic carbon during snowmelt in the Snake River near Montezuma, Colorado. Biogeochemistry 25:147-165.

Karr, J. R. 1981. Assessment of biotic integrity using fish communities. Fisheries 6:21-27.

Karr, J. R., and E. W. Chu. 1999. Restoring Life in Running Waters. Washington, D.C.: Island Press.

Kerans, B. L., and J. R. Karr. 1994. A benthic index of biotic integrity (B-IBI) for rivers of the Tennessee Valley. Ecological Applications 4:768-785.

Leavesley, G. H. 1997. The Modular Modeling System (MMS)—A modeling framework for multidisciplinary research and operational applications. Pp. 27-31 in Freeman, G. E., and A. G. Frazier (eds.) Proceedings of the Scientific Assessment and Strategy Team Workshop on Hydrology, Ecology, and Hydraulics. Volume 5 of Kelmelis, J. A. (ed.) Science for Floodplain Management into the 21st Century. Washington, D.C.: U.S. Government Printing Office.

Lenat, D. R. 1993. A biotic index for the southeastern United States: Derivation and list of tolerance values, with criteria for assigning water quality ratings. Journal of the North American Benthological Society 12:279-290.

Mallard, G. E., J. T. Armbruster, R. E. Broshears, E. J. Evenson, S. N. Luoma, P. J. Phillips, and K. R. Prince. 1999. Recommendations for Cycle II of the National Water Quality Assessment (NAWQA) Program. U.S. Geological Survey NAWQA Planning Team. U.S. Geological Survey Open-File Report 99-470. Reston, Va.: U.S. Geological Survey.

McMahon, P. B., and J. K. Bohlke. 1996. Denitrification and mixing in a stream-aquifer system: Effects on nitrate loading to surface water. Journal of Hydrology 186:105-128.

McMahon, G., and M. D. Woodside. 1997. Nutrient mass balance for the Albemarle-Pamlico Drainage Basin, North Carolina and Virginia. Journal of American Water Resources Association 33(3):73-589.

Naiman, R. J., J. J. Magnuson, D. M. McKnight, and J. A. Stanford. 1995. The Freshwater Imperative. Washington, D.C.: Island Press.

Nolan, B. T. 1998. Modeling approaches for assessing the risk of nonpoint-source contamination of ground water. U.S. Geological Survey Open-File Report 98-531. Denver, Colo.: U.S. Geological Survey.

Nolan, B. T., B. C. Ruddy, K. J. Hitt, and D. R. Helsel. 1997. Risk of nitrate in groundwaters of the United States–A national perspective. Environmental Science and Technology 31:2229-2236.

NRC (National Research Council). 1990. A Review of the USGS National Water Quality Assessment Pilot Program. Washington, D.C.: National Academy Press.

NRC. 1994. National Water Quality Assessment Program: The Challenge of National Synthesis. Washington, D.C.: National Academy Press.

NRC. 2001. Assessing the TMDL Approach to Water Quality Management. Washington, D.C.: National Academy Press.

O'Connor, D. J. 1976. The concentration of dissolved solids and river flow. Water Resources Research 12(2):279-294.

Poff, N. L., J. D. Allan, M. B. Bain, J. R. Karr, K. L. Prestegaard, B. D. Richter, R. E. Sparks, and J. C. Stromberg. 1997. The natural flow regime. Bioscience 47:769-784.

Preston, S. D., and J. W. Brakebill. 1999. Application of Spatially Referenced Regression Modeling for the Evaluation of Total Nitrogen Loading in the Chesapeake Bay Watershed. U.S. Geological Survey Water-Resources Investigations Report 99-4054. Reston, Va.: U.S. Geological Survey.

Puckett, L. J. 1999. Estimation of nitrate contamination of an agro-ecosystem outwash aquifer using a nitrogen mass-balance budget. Journal of Environmental Quality 28(6):2015-2025.

Reckhow, K. H., M. N. Beaulac, and J. T. Simpson. 1980. Modeling Phosphorus Loading and Lake Response Under Uncertainty: A Manual and Compilation of Export Coefficients. EPA-440/5-80-011. Washington, D.C.: U.S. Environmental Protection Agency.

Richter, B. D., J. V. Baumgartner, J. Powell, and D. P. Braun. 1996. A method for assessing hydrologic alteration within ecosystems. Conservation Biology 10:1163-1174.

Rupert, M. G. 1998. Probability of Detecting Atrazine and Elevated Concentrations of Nitrate ($NO_2$+$NO_3$-N) in Ground Water in the Idaho Part of the Upper Snake River Basin. U.S. Geological Survey Water-Resources Investigations Report 98-4203. Boise, Idaho: U.S. Geological Survey.

Smith, R. A., G. E. Schwarz, and R. B. Alexander. 1997. Regional interpretation of water-quality monitoring data. Water Resources Research 33(12): 2781-2798.

Spirtes, P., C. Glymour, and R. Scheines. 2001. Causation, prediction, and search. In Adaptive Computation and Machine Learning (book series). 2nd edition. Cambridge, Mass.: MIT Press.

Squillace, P. J., E. M. Thurman, and E. T. Furlong. 1993. Groundwater as a nonpoint source of atrazine and diethylatrazine in a river during base flow conditions. Water Resources Research 29(6):1719-1729.

Stalnaker, C., B. L. Lamb, J. Henricksen, K. Bovee, and J. Batholow. 1995. The Instream Flow Incremental Methodology: A Primer for IFIM. Biological Report No. 29. Fort Collins, Colo.: U.S. Department of Interior, National Biological Service.

Tesoriero, A. J., and F. D. Voss. 1997. Predicting the probability of elevated nitrate concentrations in the Puget Sound Basin: Implications for aquifer susceptibility and vulnerability. Ground Water 35:1029-1039.

Triska, F. J., V. C. Kennedy, R. J. Avanzino, G. Zellweger, and K. E. Bencala. 1989. Retention and transport of nutrients in the third-order stream in northwestern California: Hyporheic processes. Ecology 70:1893-1905.

Vervier, P., J. Gilbert, P. Marmonier, and M. J. Dole-Olivier. 1992. A perspective on the permeability of the surface freshwater-groundwater ecotone. Journal of the North American Benthological Society 11:93-102.

Vollenweider, R. A.. 1975. Input-output models with special reference to the phosphorus loading concept in limnology. Schweitz. Z. Hydrol. 37(1):53-84.

Vollenweider, R. A.. 1976. Advances in defining critical loading levels for phosphorus in lake eutrophication. Memorie dell'Istituto Italiano di Idrobiologia 33:53-83.

Winter, T. C., J. W. Harvey, O. L. Franke, and W. M. Alley. 1998. Ground Water and Surface Water: A Single Resource. U.S. Geological Survey Circular 1139. Denver, Colo.: U.S. Geological Survey.

Wright, J. F. 1995. Development and use of a system for predicting the macroinvertebrate fauna in flowing waters. Australian Journal of Ecology 20:181-197.

# 6

# Communicating NAWQA Data and Information to Users

## INTRODUCTION

The National Water Quality Assessment (NAWQA) program is first and foremost a provider of information to parties interested in water quality. While most of the NAWQA budget and effort is devoted to data collection and interpretation, it is the reporting of the program's findings that is most critical for its widespread use and the program's ultimate success. The U.S. Geological Survey (USGS) is committed to effective and timely communication of findings to managers, planners, and decision makers at all levels of government, environmental and conservation organizations, academia, industry, consulting and engineering firms, and the general public (USGS, 2001). The findings are presented in multiple formats to meet the diverse needs of the many different users, ranging from raw data to methodology, models, technical documents, journal articles, pamphlets, maps, videos, and Internet-based products.

This chapter first presents the types of information that are being produced at the national level and by individual study units. Next, methods for communicating results are assessed, including the findings of two separate outside evaluations of the effectiveness of NAWQA publications. The chapter then assesses how well NAWQA is doing to provide information that is useful to policy makers—a critical audience for its findings. The chapter concludes with recommendations on how NAWQA can improve the ways it conveys data and especially information.

## INFORMATION COMMUNICATED BY NAWQA

The NAWQA program is generating a tremendous amount of information that is of interest to researchers, resource management and regulatory agencies, and the general public. NAWQA reports, databases, and other digital products cover the entire breadth of the program, including conception, design, sampling and analysis protocols, findings, and interpretations. NAWQA information is also communicated to a wide audience, including the general public, resource managers, and other scientists. For this assessment the committee categorizes NAWQA information in two ways: (1) information coming from the national (headquarters) level and (2) information coming from individual study units.

### National

Overall, the committee finds that NAWQA is doing a good job of providing information on most aspects of the program. Information on the program has been released through USGS publications and Internet products. The USGS has produced publications on the philosophy and concepts of NAWQA (Cohen et al., 1988; Hirsch et al., 1988, 2001), providing readers with its ideas about the importance of such a national monitoring program and how it might be carried out. The USGS extensively reported on the initial design and strategy of NAWQA (Gilliom et al., 1995; Leahy and Wilber, 1991; Leahy et al., 1990) and on protocols for collecting and analyzing data (e.g., Crawford and Luoma, 1992; Cuffney et al., 1993; Fitzpatrick et al., 1998; Koterba et al., 1995; Meador et al., 1993; Shelton, 1994). NAWQA has developed and made available a number of data products, including data on groundwater, surface water, bed sediment, and animal tissue tests. As NAWQA looks ahead to the next decade (Cycle II) of the program, it has released strategic guidance and associated recommendations for where the program may go through its internal NAWQA Planning Team (NPT; Mallard et al., 1999). The NAWQA Cycle II Implementation Team (NIT) reviewed the NPT report and subsequently published *Study-Unit Design Guidelines for Cycle II of the National Water Quality Assessment (NAWQA)* (Gilliom et al., 2000). The purpose of this NIT report is to describe the design and implementation strategy for Cycle II investigations in NAWQA study units.

One shortcoming of NAWQA information is that necessary changes to the program forced by budget constraints were not fully or clearly reported during the first cycle of studies. The reasoning behind decisions to delay or discontinue study units has not been well documented, nor have changes in priorities for national synthesis studies. Using NAWQA publications solely, it can be difficult to track changes in the number of study units between 1991 and 2000. That is, of 60 study units planned for Cycle I, 2 were later merged to reduce the number of study units to 59, which often appears in NAWQA reports. However, because of budgetary constraints, eight study units that were slated for monitoring in 1997-

2001 were never initiated. Thus, Cycle I of NAWQA included a total of 51 study units and the High Plains Regional Ground Water Study that was initiated in 1999 (see Chapters 1 and 2 and Figure 1-1 for further information). Such information is available from NAWQA management but has not always been clearly conveyed.

One of the principal responsibilities of the national NAWQA office is to conduct national synthesis studies and release the findings to a wide audience, including researchers, policy makers, resource managers, and the general public. Current synthesis topics include nutrients, pesticides, volatile organics, trace elements, and stream ecology (ecological synthesis). These summary studies are an important source of information for policy makers designing national programs. The committee finds that NAWQA has done a good job to date of providing information on its national synthesis activities. The national synthesis teams provide information on program design (e.g., Gilliom et al., 1995; Halde et al., 1999; Lopes and Price, 1997; Mueller et al., 1997) and evaluation methods for comparing results across disparate study units (Gilliom et al., 1998; Nolan, 1998; Smith et al., 1997). Initial findings have been reported (Focazio et al, 1999; Goodbred et al., 1996; Larson et al., 1999; Martin et al., 1999; Mueller et al., 1995), as well as summary reports that tie study unit-level findings together (Barbash et al., 1999; Francy et al., 2000; Fuhrer et al., 1999; Nolan and Ruddy, 1996).

The USGS and NAWQA have a high-quality program for disseminating information on collected data, including field measurements and associated quality control characteristics. The NAWQA Data Warehouse (http://water.usgs.gov/nawqa/data) contains chemical, biological, and physical water quality data from all the study units, as well as site, basin, well, and network characteristics (USGS, 2001). The nationally consistent, high-quality field data provide a foundation upon which a data dissemination program builds. NAWQA has taken an active position in organizing its compiled and generated data and making it available to the public. This is the largest readily accessible water quality data set representing samples collected nationwide with consistent study design and protocols (USGS, 2001).

Public access to the NAWQA Data Warehouse is currently available through an Oracle database reached via the NAWQA Internet site. The user can make four types of queries: groundwater, surface water, mixed (ground- and surface water), and animal tissue. Information can be requested on location of samples (state, county, study unit, basin, well), chemical concentrations (500 chemical constituents), land use, daily streamflow, and groundwater levels. All the data have undergone an internal USGS quality review, and output can be produced in a variety of formats. As of May 30, 2001, data from the first 36 study units were available. Data from the 15 Cycle I study units initiated in 1997 will be added later. Further information on the Data Warehouse is presented below.

## Study Units

Each study unit is responsible for providing information on its research strategy and results. Study unit outputs describe the physical and hydrologic characteristics of the watershed, sampling design, implementation issues, results, and interpretation of findings. Many of them also provide a context for their research by identifying water quality issues that affect the entire watershed. The range of topics covered by study unit reports and other products exceeds that coming from NAWQA headquarters, simply because of the number and diversity of study unit characteristics and their individual issues, such as the impact of riparian buffers on reducing nutrient loads to groundwater in the Albemarle-Pamlico Drainages Study Unit (Spruill, 2000); the relationship between riparian cover and fish community composition in the Upper Mississippi River Basin Study Unit (Hanson, 2000); the ability of local wetlands to degrade pesticides in the Central Nebraska Basins Study Unit (Lee et al., 1995); and the impact of burley tobacco production on groundwater quality in the Upper Tennessee River Study Unit (Johnson and Connell, 2001). Several study units produced reports on the status and trends of sediment in surface waters, including the Lake Erie-Lake St. Clair Drainages Study Unit (Myers and Metzker, 2000), Apalachicola-Chatahoochee-Flint River Basins Study Unit (Frick and Buell, 1999), and Central Columbia Plateau Study Unit (Ebbert and Kim, 1998). Nutrients and sediment contained in snowmelt were studied and reported by the Upper Mississippi River Basin Study Unit (Fallon and McNellis, 2000). Data products from the Cycle I study units are available either through the central Data Warehouse or, in some cases, directly from the study units.

One issue that has been raised by researchers interested in using NAWQA study unit data is their availability and timeliness. In this regard, the committee recommends an improved procedure for releasing provisional data, particularly to collaborators, because duplicate data have been collected when researchers did not know that NAWQA already had similar data or when they could not readily access it.

## METHODS OF COMMUNICATING RESULTS

The NAWQA program conveys information in a number of ways. Most of it is in the form of published reports and journal articles. A second but very important method that is being increasingly used by NAWQA is the Internet.

## Publications

The USGS places great importance on reporting information about programs through traditional, written publications. "The written report is the principal product of the Water Resources Division of the U.S. Geological Survey. Whatever the medium for disseminating and archiving the report—paper copy, diskette, CD-ROM, or on-line—much of the speed and economy of technological advances will be wasted unless the author's initial efforts result in a technically accurate, clear, and timely document" (USGS, 1995). The NAWQA program has made an effort to ensure that publications provide useful information to the various target audiences, which include the general public, policy makers, and scientists. NAWQA has emphasized getting publications released in a timely manner.

NAWQA findings are reported in four types of USGS publications. First, Open File Reports (OFRs) are manuscripts, maps, and other materials made available for public use, generally while the material is being prepared for a more formal report. Second, Water-Resources Investigations and Reports (WRIRs) contain hydrologic information, mainly of local interest. As such, they generally are intended for quick release and contain more interpretation and analysis of data than OFR (although many OFRs also contain interpretation of data). Third, circulars contain technical or nontechnical information of popular interest, including timely administrative or scientific information. These reports generally receive broad distribution. For example, study unit summary reports and the major national synthesis reports are released as circulars. Lastly, fact sheets are very abbreviated publications that summarize research and investigations or provide details about particular USGS activities. They are used principally to get information out quickly while more detailed and formal reports are prepared. Thus, fact sheets are not intended as the sole publication for reporting findings.

A very noteworthy achievement of the NAWQA program is that as of February 26, 2001, nearly 1,000 publications and data products have been released. These include 800 reports on the findings from study unit investigations; 139

TABLE 6-1 Summary of NAWQA Publications by Type Through February 2001

| Scope and Primary Contents of Reports | Circulars | Fact Sheets | Open-File Reports | Water-Resources Investigation Reports |
|---|---|---|---|---|
| Findings from study unit investigations | 33 | 151 | 149 | 265 |
| Findings from national synthesis | 4 | 18 | 12 | 18 |
| Technical documentation of study design, field protocols, and methods comparisons | 1 | — | 20 | 6 |
| National-level general interest and outreach | 1 | 1 | 7 | — |
| Subtotal by publication type | 39 | 170 | 188 | 289 |

reports on the findings of national synthesis teams; and 34 reports documenting study design, field protocols, and methods comparisons. Of these, 39 reports have been released as NAWQA circulars, 170 as fact sheets, 188 as OFRs, and 289 as WRIRs (see Table 6-1). In addition, seven digital products have been released. The large number of WRIRs is an indication of the extent to which the USGS is interpreting data and results, often in ways that are useful for assisting policy makers and resource managers. However, all of the circulars and many of the OFRs also contain interpretation.

All USGS NAWQA publications must receive adequate technical peer review. Technical review of USGS publications may include external reviewers. This is not a requirement, but external review is commonly sought when the technical experts for a particular topic are "outside" the USGS (William Wilber, USGS, personal communication, 2001). In this regard, NAWQA has generated a large number of articles appearing in refereed professional journals. These publications provide a means for professional scrutiny of the methods and interpretations of NAWQA research. The NAWQA program is one of the first examples of an overall trend at USGS to make greater use of professional journals for reporting research. Such outlets provide a means of reaching a wider variety of audiences and a way for USGS scientists to receive greater recognition by scientific peers.

The USGS has sponsored two studies of the effectiveness of NAWQA publications. Booz-Allen & Hamilton (BAH, 1999), a management and technology consulting firm, used focus groups to obtain customer feedback on the 1991 study unit reports (all circulars), while the now-defunct National Advisory Council (see Chapters 1 and 7 for further information) conducted a more general review of NAWQA publications (Bird, 1997).

| Digital (CDs) | Conference Proceeding Papers | Journal Articles | Books, Chapters | Newsletters | Other (Professional Paper, Hydro Atlas, Thesis, Pamphlet, Web site) | Subtotal by Category |
|---|---|---|---|---|---|---|
| 4 | 64 | 112 | 4 | 8 | 10 | 800 |
| 3 | 18 | 59 | 5 | — | 2 | 139 |
| — | 3 | 3 | 1 | — | — | 34 |
| — | 4 | 9 | — | — | — | 22 |
| 7 | 89 | 183 | 10 | 8 | 12 | 995 |

The study unit reports summarize the major findings that emerged during the first three years of water quality assessments in Cycle I. The reports are intended primarily for those involved in water resource management (policy makers and managers). Ideally, the reports address concerns raised by regulators, water utility managers, industry representatives, and other scientists, engineers, public officials, and members of stakeholder groups. The focus groups employed by BAH concluded that the reports are generally well written and organized, make effective use of graphics, and are effective in reporting the results in a way that is useful to the reader (BAH, 1999). In particular, they found that information is presented in a context relevant to policy makers and resource managers. Suggested improvements included more references to other sources of information relevant to policy makers and resource managers; greater coverage of water quality issues unique to individual study units; data and documentation made available on the Internet; and more explanation of future research plans.

NAWQA's defunct National Advisory Council (NAC) convened a committee to conduct a summary evaluation of NAWQA publications (primarily fact sheets and circulars for policy makers). That committee, comprised of representatives of several federal agencies that used NAWQA information, environmental groups, and farm and industry groups, assessed the readability, relevance, and usefulness of publications for a policy audience. The NAC committee concluded that NAWQA was producing publications that generally provided sound scientific findings to water quality decision makers. It also concluded that the quality reflected efforts of NAWQA staff to make publications suitable for a policy audience. This is a noteworthy achievement given the USGS's history of writing for a more technical audience.

The NAC committee recommended that NAWQA demonstrate its commitment to further policy development and implementation by creating a series of publications geared specifically to meeting policy makers' needs for information. It also recommended that NAWQA prepare guidance for study unit and national synthesis staff to assist them in developing and writing effective publications for a policy audience. At present, NAWQA has not yet created a separate series of reports for policy makers but has made an effort to incorporate policy-relevant interpretations into its publications, as described in the next section of this chapter. The findings of the NAC committee were distributed to all Cycle I study units. The emphasis on producing timely, succinct fact sheets is partly in response to this need. Furthermore, the NAWQA Leadership Team, along with other staff, work with study unit and national synthesis personnel on an ongoing basis to provide guidance to authors on such issues (William Wilber, USGS, personal communication, 2001).

NAWQA currently has an information coordinator for overseeing the distribution of publications to appropriate audiences. Furthermore, a contractor coordinates with the National Liaison Committee and the information coordinator to assist in the development of focused publications that best serve the different audiences. Part of this coordination is helping customers use the information most effectively. However, NAWQA may want to consider the formation of a distinct information office that would provide additional resources to the important task of timely and efficient information dissemination. This office could also explore innovative strategies for getting information to Congress, resource managers, and the public.

## Internet-Based Products

The NAWQA Internet site (http://water.usgs.gov/nawqa/) is fairly easy to navigate and provides a wealth of information. The home page provides a description of the program with pertinent links to its key components, including study units, national syntheses, publications, data, and map products. A "What's New" link provides access to new publications and important updates on program-related information. The home page also provides links to other programs within the USGS that provide information on water resources, and to other federal agencies such as the U.S. Environmental Protection Agency (EPA), U.S. Department of Agriculture (USDA), and U.S. Department of the Interior.

Many of the publications produced by NAWQA headquarters, synthesis teams, and study units are available on the Web as downloadable Adobe Acrobat (PDF) files. In addition, national monitoring data sets and other information from the study units contained in the Data Warehouse are accessible from the Internet (http://water.usgs.gov/nawqa/data), along with instructions for their use. The interface for selecting data to download from the Data Warehouse is straightforward and easy to use. Data are available in a variety of formats to facilitate processing in an array of statistical or spreadsheet programs that enable users to independently evaluate questions about water quality that might not have been addressed by NAWQA personnel or otherwise reported.

However, it is important to note that the NAWQA Web site has some weaknesses. Keeping key Web pages current is a problem with any Web site, and NAWQA is not an exception. A noteworthy example of this is that the overview of the NAWQA program (http://water.usgs.gov/nawqa/NAWQA.OFR94-70.html) is very outdated, based on a 1994 report that still refers to 60 study units, rather than the 51 that were actually initiated in Cycle I. (The committee notes that this persistent oversight was recently rectified in a consistent manner.) It also does not appear that the home page provides contact information to the NAWQA Leadership Team.

As discussed previously, the USGS tends to rely on published reports to describe key components and findings of its various programs. Although these reports may be available to the general public, they are often not the most efficient way to provide information to increasing numbers of "Web surfers." The PDF format used for most documents does not allow for much interactive contact with the viewer, nor does it allow for the use of embedded links to other Internet sites.

A new feature of NAWQA's Internet site that appeared as this report was being finalized is a series of topic-driven briefing rooms describing some important findings of the program that might be relevant for policy makers or others interested in water quality issues. Although some of the links on the pages do not work, this is an effective approach for conveying important information to a wide audience. Links to published reports available on the Internet could be included to a greater degree to provide additional details for those interested in more information. An example of this approach is the USDA's Economic Research Service Web site (http://www.ers.usda.gov).

Each of the national synthesis teams has its own Web site that is accessible from the NAWQA home page. These currently include Nutrients, Pesticides, Volatile Organic Compounds (VOCs), and Trace Elements. Ecological Synthesis (still called "Aquatic Ecology") will be a future Web site. Each Web site has a very different look. The committee feels that all of the synthesis sites should include reasons for the topic (water quality issues), research strategy, important findings, bibliography, team contacts, data, and links to related sites. However, the Nutrients Web site is the only one to contain an easily accessible overview of why national research on nutrients is important. All sites contain a bibliography and a section of important findings or special topics, and all but Trace Elements contains a description of the research strategy. The Pesticides and Nutrients sites provide information on team members and how to contact them. The VOC site contains only information on the team leader, and Trace Elements provides no information on team members.

Although all of the synthesis Web sites contain links to similar or related research, some could be made more comprehensive. For example, the Nutrient Synthesis site does not contain links to EPA. Links to EPA sites covering nutrient standards development and guidance, water quality inventory, total maximum daily loads (TMDLs), and animal waste regulations would be helpful for viewers attempting to put NAWQA findings in context. The VOC site only contains links pertaining to the gasoline additive methyl *tert*-butyl ether (MTBE).

All synthesis Web sites provide links for accessing NAWQA and other data. Most provide direct links to data for constituents related to the site. However, the VOC site provides a link only to the NAWQA Data Warehouse, and the user is left to create his or her own queries for VOC data. The synthesis page also contains a link to the SPARROW (Spatially Referenced Regressions on Watershed Attributes) Web site. As discussed elsewhere in this report, SPARROW is a surface water quality model that was developed by USGS researchers outside the NAWQA Program but has become a tool often used by study units and national synthesis teams. The SPARROW Web site provides reports and articles describing the development and use of the model, examples of application at the national and watershed levels, and links to the databases used in the model. This Web site is an example of how other analytic tools developed by the NAWQA program might be displayed.

All Cycle I study units currently have or will have a Web site for conveying information about their research and findings. Each has its own look, and there is wide variation in the amount and type of information provided. The committee recommends that all (Cycle II) study unit Web sites provide easy access to the following information: description of the study unit, important water quality issues, study design, publications (as many on-line as possible), important findings, contacts, data, and links to relevant Web sites within and external to USGS-NAWQA. Although most of this information can eventually be found on all study unit sites, quite a bit of searching is often required for some of it. For example, browsing through the bibliography to find information on a particular topic is an unsatisfactory approach for providing important or basic information on a Web site. Each site should have links from the home page to important study unit information. As noted previously, this may mean putting together briefing rooms or other features that do not rely solely on published USGS documents. Guidance from the NAWQA Leadership Team on presenting information through the Internet would benefit the study units, just as guidance has resulted in more effective study unit reports.

The functionality of the Data Warehouse download interface is sometimes impaired by software problems. However, specific problems are well broadcast on the Data Warehouse main page. Moreover, highly capable and very helpful USGS staff promptly respond to help requests through the e-mail contacts posted on the site. These staff ensure that site users obtain the data they are seeking.

## POLICY RELEVANCE OF NAWQA

Information on water quality and its various dimensions is crucial for the development of effective conservation and land and water management policies in the United States. A good example is the TMDL program, where good information on water quality and sources of contaminants is vital (see Chapter 7). Policy makers need scientifically sound water quality information to answer the following questions:

- What is the nature of the problem?
- What is the extent of the problem?
- Who is affected?
- What are the causes and sources of observed water quality impairments?
- How is the situation expected to change in the future with and without action?
- Are current government and private actions adequate to address the issue or problem?
- Is more research needed before programs and regulations are developed and implemented?
- How effective have past actions been?

Despite these needs, information has not been available to answer some fundamental questions about water quality. At the time of the passage of the Clean Water Act in 1972, it was relatively easy to identify water quality problems because they were so acute. Many urban rivers were visibly impaired, the Cuyahoga even catching fire. As these problems were addressed through the Clean Water Act and other relevant regulations and policies, it became increasingly difficult to determine which waters remained threatened or impaired and what types of water quality policies were needed. The Clean Water Act set ambitious goals for national water quality, but no scientifically defensible means existed to efficiently guide the billions of dollars in public and private funds being spent on water treatment or best management practices or to determine progress toward achieving those goals. Programs designed to report on the status of the nation's water resources lacked consistency in the data collected, and sampling frameworks were not suitable for capturing the impacts of some important types of pollution, including pollution from nonpoint sources (Knopman and Smith, 1993; Leahy, 1992). In addition, the reasons or explanations for ambient conditions were not generally integrated with monitoring, so the specific causes of measured impairments could not be efficiently addressed. A 1990 General Accounting Office (GAO) report found that "important monitoring data are missing on both the scope and impacts of nonpoint source pollution and on the effectiveness of potential solutions" (GAO, 1990). A more recent GAO (2000) report echoed the same conclusion, that states have few of the data they need to effectively manage their water quality activities. This latter report also specifically identifies NAWQA as a source of information for EPA to use in evaluating national water quality conditions.

To a large degree, the NAWQA program was designed to provide policy makers with better information than was previously available. The primary goals of NAWQA—to measure the status of water quality, to identify trends, and to establish causes of observed quality—provide policy makers with the means to target programs to water resources that are impaired or threatened and to track the progress of those programs. Furthermore, NAWQA stresses quality in monitoring and interpretation. With information developed by the NAWQA program, policy makers should feel more confident about designing programs for achieving water quality goals. Not only are water quality impairments identified, but often the causes of those impairments are identified as well. The value of this information is readily seen in the ability to target policy responses to specific water quality problems. For example, rather than banning a pesticide from general use because it can impair water quality, it should be possible to target policies to protect particularly vulnerable areas. Ribaudo and Bouzaher (1994) demonstrated that such policies are more efficient than widespread bans.

Collecting scientifically sound water quality and ancillary data is important, but such information often has to be interpreted and conveyed to policy makers for it to be useful. At the time the NAWQA Program was initiated in the early

1990s, USGS scientists did not generally write reports for a policy audience. In an era of tight budgets and intense competition for funds, presenting interpretations relevant to policy makers (and budget makers) became of prime importance. As noted earlier in this chapter, NAWQA is currently providing a wealth of information on the status of water quality in the Cycle I study units through a variety of internal publications and professional outlets. While many of the publications are targeted toward a technical audience, others are targeted toward policy makers and the general public. Publications are useful for policy makers if they are issue oriented (centered around policy questions and options), direct (stating clearly what was discovered, why it is important, and how it is related to policy decisions), credible (including sufficient factual support for the findings and conclusions to be persuasive), and presented in context (providing enough information about related resource and policy factors that would be considered in a policy decision). Whereas national policy makers require a synthesis of information from local studies, augmented by specific examples of conditions in various settings around the country, local policy makers rely more on information concerning specific geographic areas.

The report *Nutrients in the Nation's Waters: Identifying Problems and Progress* (Graffy et al., 1996) aptly demonstrated that NAWQA is aware of the policy relevance of the program. This brief report outlines the water quality issue concerning nutrients, provides some geographic dimensions of the issues, and describes how NAWQA will address the issues through its research and analysis. While not providing a great deal of information, this report demonstrates to the public that the USGS and NAWQA are conducting research that can answer questions about real water quality problems affecting people's lives.

Individual study units produce research reports on a variety of issues and for a variety of audiences. The study units are using a consistent format to produce reports that provide information suitable for local policy makers and resource managers. Three representative examples are the water quality reports for the Central Columbia Plateau (Williamson et al., 1998), the Potomac River Basin (Ator et al., 1998), and the Long Island-New Jersey Coastal Drainages (Ayers et al., 2000). All three reports are very effective in summarizing the important findings in a policy context. The reports provide enough detail to establish the credibility of the science that stands behind the findings. The reports make extensive use of maps, charts, and tables to greatly enhance their effectiveness. Each report provides a summary of important findings at the beginning, making it easier for policy makers to find important information. The Long Island-New Jersey report goes a step further and includes a section devoted to implications for managing water and ecological resources. All 36 of the study units started in 1991 and 1994 of Cycle I have produced or are close to completing similar reports (USGS, 2001).

The NAWQA national synthesis reports have done an effective job of taking study unit findings, combining them with other relevant information, and presenting results in a national context. For example, the synthesis of data on nutrients and pesticides in water resources found in the report *The Quality of Our Nation's Waters: Nutrients and Pesticides* (USGS, 1999) provides a better indication of the nature of the problem, its geographic extent, and policy relevance than assessments in EPA's *Water Quality Inventory* (EPA, 2000) summarized from state reports. Policy makers and analysts seeking more detailed information can turn to other synthesis reports, including *Pesticides in Streams of the United States—Initial Results from the National Water-Quality Assessment Program* (Larson et al., 1999), *Distribution of Major Herbicides in Ground Water of the United States* (Barbash et al., 1999), and *Nutrients in Ground Water and Surface Water of the United States—An Analysis of Data Through 1992* (Mueller et al., 1995). All three of these reports provide information on the temporal dimension of water quality, which has an important bearing on the development of efficient water quality policies and is generally not reported in EPA water quality reports. The value of the national synthesis reports will increase as additional information from the second and third rounds of the Cycle I study units are added to the analyses.

As discussed in Chapter 1, an excellent example of how NAWQA data can be used to assess pollution control policies is the report *Review of Phosphorus Control Measures in the United States and Their Effects on Water Quality* (Litke, 1999). Historical data on phosphorus loads to the environment in NAWQA study units were successfully related to federal and state water quality policies for reducing phosphorus pollution from point and nonpoint sources. The report also identifies additional areas of research where NAWQA data might be used to address questions associated with the effects of pollution control policies on phosphorus.

An example of how USGS can provide policy-relevant information in response to new information is the response to the recent National Research Council (NRC, 1999) report recommendation that EPA lower the standard for arsenic in drinking water. The USGS examined data on arsenic in groundwater resources throughout the United States and identified regions where ambient (natural) concentrations of arsenic exceed possible new standards. The findings were made public through a press release, fact sheet (Welch et al., 2000), and Water-Resources Investigations Report (Focazio et al., 1999). While not an effort associated directly with NAWQA, it demonstrates that the USGS has the ability to quickly provide policy makers with information that can help them make informed decisions. These findings were provided to EPA and are being used as part of the assessment of the impact of changing the arsenic standard.

An important step in making research policy relevant is finding out what information policy makers and resource managers need to develop or implement policies. NAWQA uses its liaisons in EPA and other agencies (see Chapter 7) to see that important information is getting to the right people and to identify policy issues to which NAWQA synthesis teams and study units might contribute. Congressional and agency briefings are arranged through the NAWQA Leadership Team when important findings come to light. This approach was used effectively for the widespread findings of MTBE in groundwater.

A basic but very important question that could be asked about the nation's water quality is, Has it improved since the passage of the Clean Water Act? NAWQA is addressing important aspects of this question through its ongoing national synthesis work, but a direct discussion and explicit answer to this question could be particularly valuable to Congress, policy makers, resource managers, and the public. Even a prospective discussion of how NAWQA can or will eventually answer this type of question in Cycle II would be useful.

## NAWQA: Assisting Resource Managers

There are numerous examples of how NAWQA can or already has assisted federal and state water resource managers. As discussed earlier (see also Chapter 7), reviews by the GAO and others have noted that many states would benefit from using all available USGS information on water quality in fulfilling their reporting requirements under Section 305(b) of the Clean Water Act. Many of the differences in water quality conditions reported between states in the Water Quality Inventory are related different criteria set by each state and different methods used to characterize water quality, rather than to actual differences in water quality. At least 16 states are using NAWQA data in developing their 305(b) reports (USGS, 2001; Tim Miller, USGS, personal communication, 2001). Other states are not using their NAWQA data, but some states have very few NAWQA data to use. NAWQA was not designed to answer 305(b) questions everywhere, but it can provide important data in the study unit areas. As discussed in Chapter 7, NAWQA data can also provide direct support for requirements of the amended Safe Drinking Water Act (SDWA) and the 1996 Food Quality Protection Act. Both acts require knowledge of contaminant occurrence and distribution for risk assessments and pesticide re-registration.

An example of how NAWQA has influenced water policy is the current widespread concern over MTBE in groundwater. Synthesis reports summarizing NAWQA findings of MTBE in groundwater began coming out in 1995 (Squillace et al., 1995). These initial reports and subsequent information provided to EPA by the USGS and other agencies have resulted in the recent move by EPA to ban MTBE for use as a fuel additive.

Some states are using NAWQA data and procedures to assist their developing source water protection programs. USGS is working with the States of New Jersey and Washington to determine the risk to water supply wells from pesticide contamination (see also Chapter 7). For example, New Jersey redesigned its ambient stream monitoring network to follow the design of the Long Island-New Jersey Coastal Drainages NAWQA Study Unit. This monitoring system is being used in antidegradation analysis and TMDL planning. New Jersey is using

NAWQA data and concepts in a watershed indicators project for providing information to the planning process. New Jersey is also looking to use existing NAWQA runoff and water quality models for improving municipal land-use planning and zoning decisions. In Washington State, NAWQA information on pesticide contamination enabled the health department to identify wells with low vulnerability to contamination and to obtain waivers for monitoring required by the amended SDWA. This resulted in more than $6 million saved in monitoring costs (USGS, 2001).

NAWQA data and information on pesticides and nutrients in agricultural areas are the basis for water resource management decisions in a number of states. For example, on the basis of NAWQA findings on irrigation practices and their impacts on water quality in the Central Columbia Plateau, the local irrigation district recommended a shift from furrow to sprinkler irrigation (USGS, 2001). NAWQA findings on elevated levels of the herbicide atrazine in water supply reservoirs in the Lower Kansas River Basin were used by the state as the basis for establishing a pesticide management area in northern Kansas. NAWQA data have helped guide EPA's registration decisions on the commonly detected herbicides aldicarb, alachlor, and acetochlor and the insecticides chlorpyrifos, diazinon, and carbofuran. EPA's Office of Pesticides relies on NAWQA data to meet one of its performance goals, which states: "By 2010, detections of the 15 pesticides most frequently found in surface water in USGS NAWQA data will be reduced by 50 percent." Nevada uses groundwater data collected by NAWQA in the Nevada Basins and Range to make decisions on registering pesticides.

NAWQA's findings on how urban water resources are affected by nonpoint runoff has led to changes in the way state and local water resource managers protect urban water quality. New Jersey and the USGS are collaborating to develop a computer model based on NAWQA findings that forecasts the effect of land-use development on water quality. In the Upper Gunnison River Watershed in Colorado, local water resource managers are using NAWQA findings to determine the health of the watershed, which has seen rapid urbanization over the past 20 years. These findings are being used to guide wastewater management decisions (USGS, 2001). Lastly, NAWQA information on organochlorine compounds and trace elements in fish tissue is being used by several states, including Pennsylvania, Mississippi, New York, Ohio, Michigan, and Washington, to evaluate and establish fish consumption advisories.

## CONCLUSIONS AND RECOMMENDATIONS

The NAQWA program has generated an impressive amount of information since its inception, and has kept the public reasonably well informed of its plans and findings. Both the national synthesis teams and the individual study units are providing useful information on all facets of the program, including sampling design, implementation issues, results, and interpretations. Information is being conveyed through several types of written reports, journal articles, professional papers, digital products, the NAWQA Internet site, and an on-line Data Warehouse. Two external reviews of NAWQA publications have generally found them to be well written and to provide useful information. The USGS is taking pains to provide guidelines to NAWQA staff for producing effective reports. Similar guidance would improve the quality of information provided by national synthesis teams and study units through the Internet. NAWQA has recently started using Internet-based briefing rooms to convey findings that bear on important water quality issues. This is an effective approach for providing relevant information to policy makers and those interested in water quality issues.

Many states are using NAWQA data and findings in developing their resource management programs. This is a strong indication that NAWQA is providing valuable information to those managing water resources. However, the committee recommends that NAWQA improve on the ways it conveys information to policy makers, resource managers, and the public:

- The USGS and NAWQA have to clearly report in summary fact sheets and on the NAWQA home page changes in the scope of the program in a timely manner, along with future plans.
- As the data generated by NAWQA increase in scope and size, improvements in data access and management for outside users may be needed.
- NAWQA should ensure that data are released in a timely manner to assist researchers and resource managers. Furthermore, NAWQA should consider the formation of a distinct information office that would provide additional resources to the important task of timely and efficient information dissemination. This office could also explore innovative strategies for getting information to policy makers, resource managers, and the public.
- NAWQA should be able answer basic but very important questions about the nation's water quality such as, Has it improved since the passage of the Clean Water Act? While NAWQA's ongoing national synthesis work represents an important first step in addressing this type of question, explicit answers could be particularly valuable to Congress, policy makers, resource managers, and the public. Even a prospective discussion of how NAWQA can or will eventually answer such questions in Cycle II would be useful.
- NAWQA can provide more references (and links) to other sources of information for resource managers in both its written reports and on its Web sites.
- The NAWQA Leadership Team headquarters should continue to work with national synthesis teams and individual study units to maintain and improve the quality of written reports, to ensure that the needs of policy makers are met, and to improve the content and consistency of NAWQA Web sites.
- NAWQA should expand the use of Internet-based briefing rooms to convey important information rather than relying exclusively on electronic versions of USGS documents.

- NAWQA study units should continue to work with local resource managers to improve sampling methods, identify water quality problems, and help develop solutions to those problems.
- NAWQA should encourage staff to continue to publish NAWQA findings in refereed professional journals, where they will receive review from other scientists and broader exposure to the scientific community.

## REFERENCES

Ator, S. W., J. D. Blomquist, J. W. Brakebill, J. M. Denis, M. J. Ferrari, C. V. Miller, and H. Zappia. 1998. Water Quality in the Potomac River Basin, Maryland, Pennsylvania, Virginia, West Virginia, and the District of Columbia, 1992-96. U.S. Geological Survey Circular 1166. Reston, Va.: U.S. Geological Survey.

Ayers, M. A., J. G. Kennen, and P. E. Stackelberg. 2000. Water Quality in the Long Island-New Jersey Coastal Drainages, New York and New Jersey, 1996-98. U.S. Geological Survey Circular 1201. Reston, Va.: U.S. Geological Survey.

BAH (Booz-Allen & Hamilton, Inc.) 1999. U.S. Geological Survey National Water-Quality Assessment Program: Focus Groups for Customer Feedback on '91 Study Unit Reports. Final Report. McLean, Va.: Booz-Allen & Hamilton.

Barbash, J. E., G. P. Thelin, D. W. Kolpin, and R. J. Gilliom. 1999. Distribution of Major Herbicides in Ground Water of the United States. U.S. Geological Survey Water-Resources Investigations Report 98-4245. Sacramento, Calif.: U.S. Geological Survey.

Bird, J. C. 1997. Evaluation of National Water Quality Assessment Publications. Washington, D.C.: National Water Quality Assessment Program National Advisory Council.

Cohen, P., W. M. Alley, and W. G. Wilber. 1988. National water-quality assessment: Future directions of the U.S. Geological Survey. American Water Resources Association Bulletin 24(26):5.

Crawford, J. K., and S. N. Luoma. 1992. Guidelines for Studies of Contaminants in Biological Tissues for the National Water-Quality Assessment Program. U.S. Geological Survey Open-File Report 92-494. Lemoyne, Pa.: U.S. Geological Survey.

Cuffney, T. F., M. E. Gurtz, and M. R. Meador. 1993. Methods for collecting benthic invertebrate samples as part of the National Water-Quality Assessment Program. U.S. Geological Survey Open-File Report 93-406. Raleigh, N.C.: U.S. Geological Survey.

Ebbert, J. C., and M. H. Kim. 1998. Relation between irrigation method, sediment yields, and losses of pesticides and nitrogen. Journal of Environmental Quality 27(2):372-380.

EPA (U.S. Environmental Protection Agency). 2000. National Water Quality Inventory: 1998 Report to Congress. EPA841-R-00-001. Washington, D.C.: U.S. Environmental Protection Agency.

Fallon, J. D., and R, P. McNellis. 2000. Nutrients and Suspended Sediment in Snowmelt Runoff From Part of the Upper Mississippi River Basin, Minnesota and Wisconsin, 1997. U.S. Geological Survey Water-Resources Investigations Report 00-4165. Mounds View, Minn.: U.S. Geological Survey.

Fitzpatrick, F. A., I. R. Waite, P. J. D'Arconte, M. R. Meador, M. A. Maupin, and M. E. Gurtz. 1998. Revised Methods for Characterizing Stream Habitat in the National Water-Quality Assessment Program. U.S. Geological Survey Water-Resources Investigations Report 98-4052. Raleigh, N.C.: U.S. Geological Survey.

Focazio, M. J., A. H. Welch, S. A. Watkins, D. R. Helsel, and M. A. Horn. 1999. A Retrospective Analysis on the Occurrence of Arsenic in Ground-Water Resources of the United States and Limitations in Drinking-Water-Supply Characterization. U.S. Geological Survey Water-Resources Investigations Report 99-4279. Available online at http://co.water.usgs.gov/trace/pubs/wrir-99-4279.

Francy, D. S., D. N. Myers, and D. R. Helsel. 2000. Microbiological Monitoring for U.S. Geological Survey National Water-Quality Assessment Program. U.S. Geological Survey Water-Resources Investigations Report 00-4018. Columbus, Ohio: U.S. Geological Survey.

Frick, E. A., and G. R. Buell. 1999. Relation of land use to nutrient and suspended-sediment concentrations, loads, and yields in the upper Chattahoochee River Basin, Georgia, 1993-98. Pp. 170-179 in Hatcher, K. J. (ed.) U.S. Geological Survey reprints from proceedings of the 1999 Georgia Water Resources Conference, March 30-31. Athens, Ga.: The University of Georgia.

Fuhrer, G. J., R. J. Gilliom, P. A. Hamilton, J. L. Morace, L. H. Nowell, J. F. Rinella, J. D. Stoner, and D. A. Wentz. 1999. The Quality of Our Nation's Waters—Nutrients and Pesticides. U.S. Geological Survey Circular 1225. Reston, Va.: U.S. Geological Survey.

GAO (General Accounting Office). 1990. Greater EPA Leadership Needed to Reduce Nonpoint Source Pollution. GAO/RCED-91-10. Washington, D.C.: General Accounting Office.

GAO. 2000. Key EPA and State Decisions Limited by Inconsistent and Incomplete Data. GAO/RCED-00-54. Washington, D.C.: General Accounting Office.

Gilliom, R. J., W. M. Alley, and M. E. Gurtz. 1995. Design of the National Water-Quality Assessment Program: Occurrence and Distribution of After-Quality Conditions. U.S. Geological Survey Circular 1112. Sacramento, Calif.: U.S. Geological Survey.

Gilliom, R. J., D. K. Mueller, and L. H. Nowell. 1998. Methods for Comparing Water-Quality Conditions Among National Water-Quality Assessment Study Units, 1992-1995. U.S. Geological Survey Open-File Report 97-589. Sacramento, Calif.: U.S. Geological Survey.

Gilliom, R. J., K. Bencala, W. Bryant, C. A. Couch, N. M. Dubrovsky, L. Franke, D. Helsel, I. James, W. W. Lapham, D. Mueller, J. Stoner, M. A. Sylvester, W. G. Wilber, D. M. Wolock, and J. Zogorski. 2000. Study-Unit Design Guidelines for Cycle II of the National Water Quality Assessment (NAWQA). U.S. Geological Survey NAWQA Cycle II Implementation Team. Draft for internal review (11/22/2000). Sacramento, Calif.: U.S. Geological Survey.

Goodbred, S. L., R. J. Gilliom, T. S. Gross, N. P. Denslow, W. B. Bryant, and T. R. Schoeb. 1996. Reconnaissance of 17β-Estradiol, 11-Ketotestosterone, Vitellogenin, and Gonad Histopathology in Common Carp of United States Streams: Potential for Contaminant-Induced Endocrine Disruption. U.S. Geological Survey Open-File Report 96-627. Sacramento, Calif.: U.S. Geological Survey.

Graffy, E. A., D. R. Helsel, and D. K. Mueller. 1996. Nutrients in the Nation's Waters: Identifying Problems and Progress. U.S. Geological Survey Fact Sheet FS-218-96. Reston, Va.: U.S. Geological Survey.

Halde, M. J., G. C. Delzer, and J. S. Zogorski. 1999. Study Design and Analytical Results Used to Evaluate a Surface-Water Point Sampler for Volatile Organic Compounds. U.S. Geological Survey Open-File Report 98-651. Rapid City, S.D.: U.S. Geological Survey.

Hanson, P. E. 2000. The Relation of Fish Community Composition to Riparian Cover and Runoff Potential in the Minnesota River Basin, Minnesota and Iowa, 1997. U.S. Geological Survey Fact Sheet FS-105-00. Mounds View, Minn.: U.S. Geological Survey.

Hirsch, R. M., W. M. Alley, and W. G. Wilber. 1988. Concepts for a National Water-Quality Assessment Program. U.S. Geological Survey Circular 1021. Denver, Colo.: U.S. Geological Survey.

Hirsch, R. M., T. L. Miller, and P. A. Hamilton. 2001. Using today's science to plan for tomorrow's water policies. Environment 43(1):8-17.

Johnson, G. C., and J. F. Connell. 2001. Shallow Ground-Water Quality Adjacent to Burley Tobacco Fields in Northeastern Tennessee and Southwest Virginia. U.S. Geological Survey Water-Resources Investigations Report 01-4009. Nashville, Tenn.: U.S. Geological Survey.

Knopman, D., and R. Smith. 1993. 20 Years of the Clean Water Act. Environment 35(1):184-194.

Koterba, M. T., F. D. Wilde, and W. W. Lapham. 1995. Ground-Water Data-Collection Protocols and Procedures for the National Water-Quality Assessment Program: Collection and Documentation of Water-Quality Samples and Related Data. U.S. Geological Survey Open-File Report 95-399. Reston, Va.: U.S. Geological Survey.

Larson, S. J., R. J. Gilliom, and P. D. Capel. 1999. Pesticides in Streams of the United States—Initial Results from the National Water-Quality Assessment Program. U.S. Geological Survey Water-Resources Investigations Report 98-4222. Sacramento, Calif.: U.S. Geological Survey.

Leahy, P. P., J. S. Rosenshein, and D. S. Knopman. 1990. Implementation plan for the National Water-Quality Assessment Program: U.S. Geological Survey Open-File Report 90-174. Reston, Va.: U.S. Geological Survey.

Leahy, P. P. 1992. Consistent data on water quality—it's long overdue. Geotimes 27(12):5.

Leahy, P. P., and W. G. Wilber. 1991. National Water-Quality Assessment Program. U.S. Geological Survey Open-File Report 91-54. Reston, Va.: U.S. Geological Survey.

Lee, K. E., D. G. Huggins, and E. M. Thurman. 1995. Effects of hydrophyte community structure on atrazine and alachlor degradation in wetlands. Pp. 525-538 in Campbell, K. L. (ed.) Versatility of Wetlands in the Agricultural Landscape. St. Joseph, Mich.: American Society of Agricultural Engineers.

Litke, D. W. 1999. Review of Phosphorus Control Measures in the United States and Their Effects on Water Quality. U.S. Geological Survey Water-Resources Investigations Report 99-4007. Denver, Colo.: U.S. Geological Survey.

Lopes, T. J., and C. V. Price. 1997. Study Plan for Urban Stream Indicator Sites of the National Water-Quality Assessment Program. U.S. Geological Survey Open-File Report 97-25. Rapid City, S.D.: U.S. Geological Survey.

Mallard, G. E., J. T. Armbruster, R. E. Broshears, E. J. Evenson, S. N. Luoma, P. J. Phillips, and K. R. Prince. 1999. Recommendations for Cycle II of the National Water Quality Assessment (NAWQA) Program. U.S. Geological Survey NAWQA Planning Team. U.S. Geological Survey Open-File Report 99-470. Reston, Va.: U.S. Geological Survey.

Martin, J. D., R. J. Gilliom, and T. L. Schertz. 1999. Summary and Evaluation of Pesticides in Field Blanks Collected for the National Water-Quality Assessment Program, 1992-95. U.S. Geological Survey Open-File Report 98-412. Indianapolis, Ind.: U.S. Geological Survey.

Meador, M. R., T. F. Cuffney, and M. E. Gurtz. 1993. Methods for Sampling Fish Communities as Part of the National Water-Quality Assessment Program. U.S. Geological Survey Open-File Report 93-104. Raleigh, N.C.: U.S. Geological Survey.

Mueller, D. K., P. A. Hamilton, D. R. Helsel, K. J. Hitt, and B. C. Ruddy. 1995. Nutrients in Ground Water and Surface Water of the United States—An Analysis of Data Through 1992. U.S. Geological Survey Water-Resources Investigations Report 95-4031. Denver, Colo.: U.S. Geological Survey.

Mueller, D. K., J. D. Martin, and T. J. Lopes. 1997. Quality-Control Design for Surface-Water Sampling in the National Water-Quality Assessment Program. U.S. Geological Survey Open-File Report 97-223. Denver, Colo.: U.S. Geological Survey.

Myers, D. N., and K. D. Metzer. 2000. Status and Trends in Suspended-Sediment Discharges, Soil Erosion, and Conservation Tillage in the Maumee River Basin—Ohio, Michigan, and Indiana in Cooperation with the U.S. Army Corps of Engineers and the U.S. Department of Agriculture, Natural Resources Conservation Service. U.S. Geological Water-Resources Investigations Report 00-4091. Columbus, Ohio: U.S. Department of the Interior.

Nolan, B. T. 1998. Modeling Approaches for Assessing the Risk of Nonpoint-Source Contamination of Ground Water. U.S. Geological Survey Open-File Report 98-531. Denver, Colo.: U.S. Geological Survey.

Nolan, B. T., and B. C. Ruddy. 1996. Nitrate in Ground Waters of the United States—Assessing the Risk. U.S. Geological Survey Fact Sheet 092-96. Reston, Va.: U.S. Geological Survey.

NRC (National Research Council). 1999. Arsenic in Drinking Water. Washington, D.C.: National Academy Press.

Ribaudo, M. O., and A. Bouzaher. 1994. Atrazine: Environmental Characteristics and Economics of Management. Agricultural Economic Report 699. Washington, D.C.: U.S. Department of Agriculture, Economic Research Service.

Shelton, L. R. 1994. Field Guide for Collecting and Processing Stream-Water Samples for the National Water-Quality Assessment Program. U.S. Geological Survey Open-File Report 94-455. Sacramento, Calif.: U.S. Geological Survey.

Smith, R. A., G. E. Schwarz, and R. B. Alexander. 1997. Regional interpretation of water-quality monitoring data. Water Resources Research 33(12):2781-2798.

Spruill, T. B. 2000. Statistical evaluation of effects of riparian buffers on nitrate and ground water quality. Journal of Environmental Quality 29(5):1523-1538.

Squillace, P. J., J. S. Zogorski, W. G. Wilber, and C. V. Price. 1995. A Preliminary Assessment of the Occurrence and Possible Sources of MTBE in Ground Water of the United States, 1993-1994. U.S. Geological Survey Open-File Report 95-456. Rapid City, S.D.: U.S. Geological Survey.

USGS (U.S. Geological Survey). 1995. Guidelines for Writing Hydrologic Reports. U.S. Geological Survey Fact Sheet 217-95. Reston, Va.: U.S. Geological Survey. Available online at http://water.usgs.gov/wid/FS_217-95/guide.html.

USGS. 1999. The Quality of Our Nation's Water: Nutrients and Pesticides. U.S. Geological Survey Circular 1225. Reston, Va.: U.S. Geological Survey.

USGS. 2001. The National Water-Quality Assessment Program: Informing Water-Resource Management and Protection Decisions. Reston, Va.: U.S. Geological Survey. Available online at http://water.usgs.gov/nawqa/docs/ xrel/external.relevance.pdf.

Welch, A. H., S. A. Watkins, D. R. Helsel, and M. J. Focazio. 2000. Arsenic in Ground-Water Resources of the United States. U.S. Geological Survey Fact Sheet FS-063-00. Denver, Colo.: U.S. Geological Survey.

Williamson, A. K., M. D. Munn, S. J. Ryker, R. J. Wagner, J. C. Ebbert, and A. M. Vanderpool. 1998. Water Quality in the Central Columbia Plateau Washington and Idaho. U.S. Geological Survey Circular 1144. Tacoma, Wash.: U.S. Geological Survey.

# 7

# Cooperation and Coordination Issues

## INTRODUCTION

The national scope of the National Water Quality (NAWQA) Program and its potential to provide a nationwide perspective on the status, trends, and understanding of factors that affect water quality, have made it a focal point within the Water Resources Division (WRD) of the U.S. Geological Survey (USGS). Indeed, many local, state, and even federal agencies and organizations, many of which have not worked with the USGS in the past, now regularly promote the use of NAWQA data and information (though sometimes without a full understanding of their inherent limits and availability). Furthermore, the increased use and visibility of NAWQA data and information often occur in conjunction with attempts to influence the design or to cooperate to broaden NAWQA's coverage. With a program of such scope, cooperation, coordination, and real collaboration with external agencies and organizations (e.g., related to budgets and staffing) should be priorities to optimize the massive data collection, as well as data use and interpretation, efforts. Indeed, every preceding chapter of this report notes examples of such cooperative efforts. The committee applauds the USGS for its work in this arena. In the committee's view, NAWQA program staff have done an excellent job of establishing cooperative relationships within USGS and with external programs. These efforts have strengthened NAWQA and have improved the visibility and viability of the USGS as a whole. Cooperation has costs, however, particularly in staff time to keep effective communications in place. Also, while cooperative efforts are valued, NAWQA cannot and should not attempt to be all things to all people—or agencies.

NAWQA must stay firm in its design to meet its national goals and should not change critical design plans to meet the diverse needs of the many federal, state, and local agencies that seek to participate in the program or utilize its data and information. Thus, NAWQA must uphold its careful balancing act to uphold its design principles while finding new ways to collaborate that build and improve the program as it enters Cycle II. Perhaps more importantly, such collaboration should strive to improve and strengthen other water-related programs to enhance the total knowledge of the nation's water resources. To do this, other agencies that want to utilize NAWQA or to coordinate programs with NAWQA also have a responsibility to fully collaborate with the program (i.e., to give, not just take). As large as NAWQA is, its program resources are often too constrained to fully meet its national goals. Continued flat budgets or budget cuts will not allow NAWQA to meet its goals or to provide the information that Congress and other agencies desire without continued design changes and cutbacks. The significant scaling back of study units in Cycle II discussed in Chapter 2 is but one example. Further programmatic and design changes that can improve the efficiency and cost-effectiveness of NAWQA in Cycle II are certainly warranted and should continue to be developed. However, providing a national perspective on the nation's water quality requires adequate support. While Congress must recognize this, other agencies also have to contribute their due support, with staffing, fiscal, or in-kind support where possible, to help cover their unique needs and requests. The USGS also should try to ensure reasonable overhead to keep transaction costs affordable for cooperators.

A potentially sensitive issue in cooperative programs can be ensuring that full and proper credit is given where credit is due. Often the largest or most visible program in a cooperative effort receives the majority of credit for its accomplishments, whether warranted or not. All cooperators need to be aware of this common and unfortunate situation to help ensure that ample credit is provided and directed to appropriate parties. To some observers, NAWQA has become synonymous with USGS water resources programs, particularly among those new to using USGS information. This cannot help but strain important internal relationships. NAWQA must continue to recognize and share credit with its internal and external partners.

This chapter outlines several pertinent examples of cooperative efforts that are taking place within NAWQA. Through these examples, the committee hopes to address the various cooperation and coordination concerns and issues raised at committee meetings or during interviews and meetings with USGS-NAWQA personnel and others with a vested interest in NAWQA data and information. These examples address both benefits and problems for the program, illustrate typical management challenges, and hopefully, identify some new or expanded opportunities for cooperation and collaboration.

## NAWQA LIAISON COMMITTEES

A particularly successful logistical component of NAWQA that began with its pilot studies was the development of local and national coordination and advisory groups. Establishment of individual study unit liaison committees as a component of NAWQA was a recognition by the USGS of the importance of obtaining local information and perspectives in water resource issues. In this regard, every water body and watershed within the United States has some sort of local constituency that should be included in the design, planning, and execution of such studies. The study unit liaison committees are able to make valuable contributions to NAWQA, assisting with local coordination, providing local knowledge and insight of water resources problems, and bringing other expertise into the design and review process. These groups are also some of the most important consumers of the detailed information generated by NAWQA at the study unit level (see Chapter 1 for a discussion of the typical membership of study unit liaison committees). It is essential that these committees be continued in Cycle II as currently planned. Particularly as Cycle II places more emphasis on cause-and-effect studies, local and regional expertise may become even more important. As noted in previous National Research Council (NRC) reports, NAWQA has in the past used the liaison committees more to obtain technical information than local policy information. The committee feels that this represents an area that can be better exploited by NAWQA personnel. NAWQA leadership has also internally developed a work group to assess the local liaison committee process and make recommendations for its improvement. Based on this committee's observations, it would seem that the success of the local liaison committees has been highly variable. In some study units, the only real activities have been to hold infrequent meetings to present monitoring status and results, whereas in others, valuable dialogue has taken place, resulting in local, jointly funded projects.

The national NAWQA Advisory Council was formed in 1991 as a panel of federal scientists representing agencies having an interest in water quality issues. The purpose of the council was to advise the USGS on plans and activities related to the relatively new NAWQA Program and the effectiveness of NAWQA in meeting national needs. The council sought to help ensure both the use of the best and most current scientific methods and the national relevance of the program's findings. The council also served as a forum for the exchange of information about water quality-related activities being conducted by other organizations and identified opportunities to collaborate, transfer technology, and share information.

In 1993, the structure of the council was changed to include nonfederal representatives from the Advisory Committee on Water Data for Public Use (ACWDPU). More specifically, the NAWQA Advisory Council was made a subcommittee of the ACWDPU that operated under the rules established by the

Federal Advisory Committee Act. The Secretary of the Interior designates the membership of the ACWDPU. Its members are regional, state, and national organizations having a wide-ranging interest in water resources, including Native American groups, professional and technical societies, public interest groups, private industry, and the academic community. All member organizations were invited to designate a representative to serve on the NAWQA Advisory Council. The council met at least once per year from 1991 through 1997 (12 meetings in all).

To improve the interaction between the USGS and external parties with an interest in the success of the NAWQA program, the national liaison structure was changed again in 1999. The NAWQA Advisory Council was replaced by the NAWQA National Liaison Committee (NLC). This new committee is convened as a subcommittee of the Interagency Advisory Council on Water Information and also operates under Federal Advisory Committee Act requirements. The new NLC is smaller than the council it replaced and includes several member organizations that use scientific information to make water policy and management decisions.

In the opening meeting of the NLC, USGS staff noted that it views the new liaison committee as a catalyst for the scientifically informed examination of water quality issues and policy decisions. Benefits are anticipated to include enhanced application of NAWQA information in the policy context, identification and establishment of collaborative projects to benefit NAWQA and its liaison participants, and continuation of an open forum for discussing the consequences of scientific findings for water policy and policy options on water quality. Membership on the committee is expected to fluctuate over time as interested organizations' priorities change and as NAWQA's focus evolves.

The NLC will meet as needed for specifically identified purposes. NAWQA staff will also work individually with members and participants to explore opportunities for mutually beneficial work and to conduct joint activities. The first two meetings of the liaison committee were held in 2000 in Washington, D.C. At these meetings, the USGS sought feedback on NAWQA's urban water quality activities and on proposed themes for Cycle II.

The committee recommends that the local and national liaison committees be continued as planned in Cycle II. NAWQA should also consider providing training as well as further information sharing ("lessons learned") for study unit teams to make the local liaison efforts more beneficial for USGS and local users.

## USGS DISTRICT PROGRAMS

The primary operational unit of the USGS water resources field programs (e.g., stream gaging) has been the local district offices, generally organized at the state level. The NAWQA program's field operations also operate in and through these district offices. While NAWQA has brought new resources and opportunities to the districts, it also provides new stresses and management challenges. Overall, the relationship between district offices and NAWQA has been viewed as efficient and beneficial because it has brought resources to the local level. In addition, this relationship retains the details of management and staffing coordination at the local level, where such decisions can most efficiently be made. Such collaborative efforts, even though internal, require continued attention in Cycle II. Based on the committee's investigations and deliberations during this study, there are two key areas of stress in which continuing dialogue is needed. The first is the management problems created by the fluctuation in resources and staffing that occurs between the high-intensity and low-intensity phases of NAWQA. The high-intensity phases require more staff and resources, but there may not be the resources (or work) to support this level of staffing in the same district during the low-intensity phase, thus requiring that staff move, or that other adjustments be made. Also, the planning cycles and budget decisions are sometimes not in phase, creating many uncertainties and anxieties. A second concern is that the district-level staff fully appreciate the national goals of NAWQA, and likewise that NAWQA study unit personnel listen to design and implementation concerns from the local expertise. As noted previously, although NAWQA's design cannot be altered to meet every local need, local concerns must at least be acknowledged. The continued success of NAWQA must involve managing the "creative tension" of the joint effort between the districts and the national program staff.

## OTHER USGS WATER RESOURCES RESEARCH PROGRAMS

The USGS must continue to foster a productive symbiosis among field monitoring and research programs such as NAWQA, the National Research Program (NRP), and the Toxic Substances Hydrology ("Toxics") Program. To date, the interaction and collaboration among the NRP, the Toxics Program, and NAWQA have been clearly valuable for all three programs. Undoubtedly, the collaboration creates some amount of stress and tension because these programs must also, to some extent, compete for recognition and resources.

The NRP and the Toxics Program have provided NAWQA with new methods and tools for investigations, as well as approaches to study designs. In turn, NAWQA has provided an avenue for national-scale testing for contaminants and testing of concepts from smaller-scale investigations. Internal funding has crossed between programs, and in some cases, staff have been shared. This internal interaction should clearly be continued and strengthened where possible and appropriate. Particularly, the increased focus in Cycle II on "understanding" (see Chapter 5) the major factors that affect water quality conditions and trends would seem to call for stronger interaction. Some logical examples of useful collaboration have become apparent during the committee's investigations. For example, can the NRP help NAWQA investigate and understand the details of surface water-groundwater interaction relevant to water quality and unravel details of processes in the hyporheic zone? Further, can methods be developed and routinely applied or tested within NAWQA to help develop knowledge in this important area and provide data to construct better models for future work? If the Toxics Program can validate sampling and analytical methods for assessing the occurrence of pharmaceuticals in water supplies, can these methods be applied in NAWQA to begin to assess their importance nationally?

Increased collaboration among NAWQA, NRP, and the Toxics Program may also provide opportunities to help resolve management and staffing concerns that arise because of the timing and cycling of NAWQA efforts within and among USGS districts. Also, with an increased emphasis on studies concerning the understanding goal in Cycle II, expanded interaction and involvement with states and universities may also prove beneficial. These collaborative activities may be facilitated by furthering the existing relationship among NAWQA and various state Water Resources Research Institutes.

### Atmospheric Deposition (National Atmospheric Deposition Program)

The National Trends Network (NTN) of the National Atmospheric Deposition Program is also operated by the USGS's WRD, and sometimes NAWQA staff support these activities. The networks and programs were designed for different purposes, at different times, and were not designed to be coincident. The major connection, however, is that the NAWQA program utilizes data collected under the NTN. In particular, a task of most study units is to develop a nutrient budget (and mass balance) to help understand the major factors that affect nutrient loading and transport. NTN data serve as the major source of information on atmospheric loading for such constituents and have been used by NAWQA for local and national nitrogen assessments (e.g., Puckett, 1994). A limited number of NTN sites have also been sampled for mercury over time, and such data could be valuable for Cycle II assessments.

### Relationship Between NAWQA and NASQAN

The National Stream Quality Accounting Network (NASQAN) program began in 1973. Prior to NAWQA, NASQAN served as the sole USGS network for evaluating stream water quality at the national scale. In its original 1973 program, NASQAN had a gage at the downstream end of every accounting unit (i.e., major watershed) and included 500 stations by the late 1970s. However, these data were not widely used, and the program steadily shrank with the growing success of NAWQA. In 1996, the network was significantly redesigned using the Kendall's tau statistical method to justify low-frequency measurements. This decision did not accommodate physical aspects of water quality variations, only statistical aspects. Many NASQAN stations have become NAWQA surface water sites. The NASQAN program now focuses exclusively on very large river basins.

There are currently 40 monitoring stations scattered across 4 major river basins: 17 in the Mississippi basin; 9 in the Rio Grande; 8 in the Colorado; and 6 in the Columbia. Hooper et al. (1996) describes the NASQAN program redesign in greater detail.

The current objectives of NASQAN are to characterize the concentrations and flux of sediment and chemicals in the nation's largest river basins, to determine regional source areas for these materials, and to assess the effect of human influences on observed concentrations and flux (Hooper et al., 1996). NASQAN complements NAWQA by adding consistent measurements of concentrations and transport of constituents on the main stem of large rivers downstream of NAWQA study units. NASQAN further complements NAWQA by providing additional relational cause-and-effect data for interpreting water quality conditions (such as relationships among land use and contaminants) and trends in the largest rivers. NASQAN data could be used to test and verify regional inferences about water quality developed in NAWQA studies. Such comparisons might potentially help to evaluate how confidently NAWQA data can be extended to unmeasured areas and regional issues (see discussion on representativeness and extrapolation in Chapter 2). Although this complementary design can be effective, some coverage has been lost. Some surface water stations were dropped in basins intermediate in size between NAWQA and NASQAN, and coverage of coastal watersheds was lost except where NAWQA may include them.

In some cases, the same location may be sampled by both programs. Ongoing coordination will be necessary, from shared station operation, through sampling protocols and ultimate data analysis. NAWQA program documents (Mallard et al., 1999) suggest that a standing committee will be formed to ensure communication and coordination between the programs. The committee recommends that such a coordinating body be created.

## COOPERATION WITH OTHER AGENCIES

The USGS, in part through NAWQA, has developed some very important formal liaisons and cooperative programs with other federal agencies, particularly several national offices of the U.S. Environmental Protection Agency (EPA). For example, USGS staff liaisons are in residence in the Office of Water and working in support of the Safe Drinking Water Act (SDWA) Amendments of 1996 (Office of Ground Water and Drinking Water, OGWDW); the Clean Water Act (Office of Wetlands, Oceans, and Watersheds); and the development of water quality standards and criteria (Office of Science and Technology). Other USGS staffs are in the Office of Pesticide Programs (OPP) helping to provide information and technical support about pesticide occurrence in water, as well as fate and transport perspectives. In general, USGS staff liaisons provide EPA with important scientific perspective in technical approaches to water quality assessments, development of regional nutrient criteria, and the identification of emerging contaminants that are, or should be, included on EPA's Drinking Water Contaminant Candidate List (CCL). Such liaison work has provided important technical expertise to EPA and has also provided the USGS with important perspectives on EPA's water-related responsibilities and corresponding information needs. Further, this has resulted in EPA's providing funding to USGS to help jointly underwrite projects complementary to NAWQA, sometimes helping to fill funding gaps in NAWQA.

For example, OPP has helped to support development of approaches to model and extrapolate NAWQA and NASQAN data to national estimates of pesticide occurrence in all watersheds (Larson and Gilliom, 2001). Further, OPP and USGS are jointly developing a study to sample and assess pesticides in water supply reservoirs, lakes, and finished drinking water. This study will help to address the shortcoming of NAWQA's largely excluding lakes and reservoirs from monitoring as discussed in Chapter 2. The USGS and EPA cooperatively assessed the occurrence of arsenic and methyl *tert*-butyl ether (MTBE) in water to evaluate the need for regulatory action (Focazio et al., 1999; USGS, 2001; Welch et al., 2000; Zogorski et al., 1997).

Development of the new Cycle II status theme on drinking water resources (see Chapter 3) is undoubtedly in part an outgrowth of the direct interaction of USGS water resources staff and EPA's OGWDW programs. In particular, it is hoped that EPA's knowledge of drinking water systems has helped USGS in the design for implementation of this theme. As a further component, NAWQA and EPA are developing a joint effort for USGS involvement in studies related to the new Unregulated Contaminant Monitoring Regulation (UCMR) program required under the amended SDWA. The USGS also has developed a new Drinking Water Initiative that crosscuts NAWQA and other programs, involving state and local cooperative programs to evaluate source waters and public water system (PWS) vulnerability and other particular PWS problems. In addition, the USGS is working with EPA to improve the River Reach data files that are widely used for cataloging water quality data.

## Coordination to Identify Emerging Contaminants

The USGS and NAWQA, in particular through staff liaisons, have worked with EPA to identify emerging contaminants that warrant consideration for regulation or monitoring to protect public health and maintain water quality. In this regard, EPA has found NAWQA data to be very useful for considering whether chemical contaminants included on the 1998 CCL represent a significant potential threat to drinking water supplies (e.g., EPA, 2001). The long list of contaminants being monitored in NAWQA is already providing information for consideration for future CCL lists, as well. In its interaction with EPA and other interested groups to assess new and emerging drinking water contaminants, NAWQA has also been responsive in reviewing whether such contaminants can be included in its monitoring. For example, NAWQA is considering adding some contaminants that EPA has begun monitoring under the new UCMR (William Wilber, USGS, personal communication, 2001).

In practice, NAWQA has taken the lead in gathering such data. Many of the analytical methods used in NAWQA are multiresidue methods (i.e., they can detect a broad spectrum of contaminants in a single sample). For example, USGS has broadened the monitoring of pesticides to include many of the feasible compounds that can be derived from a single sample, including many degradation products of parent pesticide compounds. Such environmental degradates have been of growing interest to EPA and the public health research community in recent years, and future data from NAWQA on the range of their occurrence will be critical to evaluate their potential importance. Also, even nontarget compounds that may appear in samples are periodically identified in the NAWQA program (e.g., using computerized searches of mass spectral libraries). For those nontarget compounds that appear frequently, efforts have been made to include them as new targets, where appropriate (Miller and Wilber, 1999). Through just such a process, in 1991, the gasoline additive MTBE was discovered to be widely occurring in NAWQA and Toxics Program samples, and it was subsequently added to the USGS volatile organic compound (VOC) analyte list (see also Chapter 1). These data have played an important role in EPA's and several states' deciding to take a thorough review of MTBE's impact on drinking water supplies and public heath. Furthermore, EPA and USGS have recently entered into a jointly funded program to conduct more detailed monitoring for MTBE in some critical states where it is suspected to be a greater problem (USGS, 2001).

## Environmental Monitoring and Assessment Program

As discussed in Chapter 2, the EPA's Environmental Monitoring and Assessment Program (EMAP) was initiated about the same time as NAWQA and was intended to monitor the status and trends in the nation's ecological resources. This included some parallel goals concerned with the quality of surface and groundwater. There is no formal relationship between NAWQA and EMAP, in large measure because EMAP has not been fully developed or funded. Although a strategy for EMAP has been developed, there is no national implementation plan with which NAWQA can interface. Rather, most current EMAP activities are in the research mode and are regional in nature (commonly referred to as RE-MAP) (e.g., Hellkamp et al., 2000; Hess et al., 2000; Jones et al., 1997). However, NAWQA staff should consider coordinating with these future regional teams and activities as warranted.

## Coordination with Estuary Programs

NAWQA, as discussed earlier (Chapters 2 and 3), does not assess coastal waters that include estuaries, partly because the National Oceanic and Atmospheric Administration (NOAA) has responsibilities for estuaries and has its assessment program. EPA, in conjunction with NOAA, also has established the National Estuary Program. NAWQA does collaborate with these programs (e.g., the Chesapeake Bay Program), however, and many NAWQA study units are relevant to estuarine water quality, being the upstream contributors of water and contaminants to the estuary. Thus, NAWQA monitoring data are of importance to estuarine studies. As noted previously in Chapter 3, NAWQA data were used to establish loadings to a number of areas in the recent assessment of nutrient enrichment in the nation's estuaries. For these collaborative studies, coordination is established at the local and regional levels as related to the particular relevant study units. In some cases, NAWQA and the USGS have contributed special resources to the estuary studies, as noted in Box 7-1.

## NAWQA AND THE STATES

States are increasingly becoming important users of NAWQA data. As discussed in Chapter 6, reports from the U.S. General Accounting Office (GAO, 2000) and others note that states could benefit from using NAWQA data in two required programs, the source water assessments under the SDWA and the water quality assessments ("305[b]" programs) under the Clean Water Act (CWA). States can also interact directly with the administration of NAWQA through the study unit liaison committees and can utilize NAWQA data and cooperation in many other programs ranging from nonpoint source programs to state pesticide management programs.

More than 30 states are using USGS information for source water assessments, and at least 18 states use NAWQA data for water quality assessments (USGS, 2001). Examples are discussed below.

### Source Water Assessment Programs

Several states have collaborated with USGS district office and/or NAWQA personnel to assess the vulnerability of their drinking water sources and public drinking water supply systems as required by Sections 1428 and 1453 of the SDWA Amendments of 1996 (see Box 7-2). More specifically, the amended SDWA requires states to develop, submit to EPA, and after approval, implement Source Water Assessment Programs (SWAPs). Under the SWAP, the states are required to complete assessments for all public water systems. A SWAP consists of three elements: (1) delineation of source water protection areas, (2) inventories of certain contamination sources, and (3) determination of susceptibility of PWSs

## BOX 7-1
## An Estuarine Model: DSM2 for the Sacramento-San Joaquin Delta, California

The Sacramento-San Joaquin Delta in northern California receives about 47 percent of California's annual runoff mainly from the Sacramento River to the north and the San Joaquin River to the south. Waters exported from the delta serve as drinking water for more than 22 million people and about 45 percent of the irrigation needs in the state (CADWR, 1993). Outflow from the delta during high flow produces flushing action in San Francisco Bay and during low flow helps repel seawater intrusion. Delta waters support numerous fish and wildlife, some of them threatened and endangered species. There are more than 1,000 miles of levees protecting 67 islands from inundation, with a total surface area of about 690,000 acres. The management of delta waters for the environment and exportation is a source of continuing controversy. CALFED Bay-Delta Program, a joint state-federal effort, is seeking alternative measures to restore the delta habitat and provide water supply. CALFED is utilizing hydraulic simulation models to evaluate alternative management options.

The State of California's Department of Water Resources (DWR), a major participant in CALFED, has developed Delta Simulation Model II (DSM2) for the Sacramento-San Joaquin Delta. (DSM2 is copyrighted by DWR and is available to the public [http://wwwdelmod.water.ca.gov/simulations/dsm2/dsm2.html].) DSM2 is comprised of three modules: DSM2-HYDRO for hydrodynamics, DSM2-QUAL for water quality, and DSM2-PTM for particle tracking. DSM2 may be interfaced with DICU (see below), which computes consumptive use by crops in the Delta islands and the quality of return flow, and the THM module, which estimates THM (trihalomethane) formation potential and speciation.

DSM2-HYDRO and its salinity submodel can calculate water stages, flows, velocities and salinity in the river and estuary. DSM2-QUAL can simulate the reactivity and transport of carbonaceous biochemical oxygen demand and dissolved oxygen, nitrogen species (organic N, $NH_3$, $NO_2$, and $NO_3$), organic and dissolved phosphorus, and algal growth and respiration. DSM2-PTM simulates the transport and fate of individual particles in three dimensions and has been used to simulate, for example, the transport and mortality of eggs and larvae of striped bass. The DICU model considers water balance and water salinity on 142 delta island subareas with about 250 channel irrigation diversion nodes and 200 irrigation drainage nodes. The THM module simulates DOC (dissolved organic carbon) and THM species, which are disinfection by-products formed from chlorination of drinking waters. Because of the prominent presence of bromine in seawater intruding into the delta, bromine distribution factors are also incorporated into the THM formation process.

NAQWA has study units in two basins that serve as source water and also as the source of some of the nonpoint source pollutant inputs to the delta: the San Joaquin-Tulare River Basin and the Sacramento River Basin (combined into one study unit for Cycle II; see Chapter 2). NAWQA is conducting research on two constituents of concern in the delta: diazinon and methyl mercury. Diazinon is the most commonly used dormant insecticide spray to control wood boring insects in fruit trees. About 1 million pounds of diazinon is applied annually in January and February to about 0.5 million acres of orchards in the San Joaquin and Sacramento Valleys (Domagalski et al., 1998). USGS monitoring has detected concentrations of diazinon exceeding aquatic toxicity reference levels (0.35 µg/L) during winter storm runoffs for several years in the San Joaquin and Sacramento Rivers (Dubrovsky et al., 1998). The potential toxicity of diazinon to aquatic biota is being evaluated with seven-day bioassays using water fleas (*Ceriodaphnia dubia* ). NAWQA is also monitoring elevated concentrations of trace metals, especially mercury from drainage waters from abandoned mines and tailings (e.g., quicksilver, copper, zinc, gold) in the Sacramento River watershed as well as sediment mercury laid down in the river system during hydraulic mining activities in the Gold Rush period (Domagalski et al., 1998). Of primary concern is the presence of methyl mercury in the river system and potential resulting toxicity to human health and wildlife. NAWQA has monitored for total mercury and methyl mercury in Cache Creek in the Coast Range and several tributaries in the Sierra Nevada, both in the Sacramento River watershed. Currently, the USGS and NAWQA are investigating mercury occurrences, methylation and demethylation, transport, and bioaccumulation in the aquatic food chain (e.g., Alpers et al., 1999, 2000). NAWQA personnel are interacting with CALFED in a number of programs such as water quality and ecosystem restoration. Because of the contentious water export and water quality issues, the newly merged Cycle II NAQWA study unit might also interface directly with DSM2 and help extend DWR's DSM2 to pesticides (e.g., diazinon) and metals (e.g., methyl mercury).

**BOX 7-2**
**New Jersey as a Case Study**

The state of New Jersey is working closely with local USGS personnel in developing its source water assessment program (SWAP). New Jersey's SWAP (http://www.state.nj.us/dep/watersupply/swap2.htm) consists of six activities, the first of which is to identify current and future threats to the water supply. It involves the use of data collected under NAWQA, as well as other programs such as those conducted by the New Jersey Bureau of Safe Drinking Water. This activity is divided into three steps: (1) conducting a susceptibility assessment, (2) evaluating existing finished water quality to check the accuracy of the susceptibility assessment, and (3) determining the type of treatment in place at the water treatment facility. The susceptibility assessment is a predictive tool for managing protection, treatment, and monitoring of the source. It is based on the sensitivity of the drinking water source to contamination from land-use activities and the intensity of use of the contaminants within the delineated area.

In order to assess the susceptibility, it is necessary to have the following information: (1) accurate locations of each source of drinking water and a delineation of the area of concern around the water source, (2) an evaluation of the intensity of contaminant use or occurrence in the delineated area, and (3) an analysis of the inherent hydrogeological sensitivity of the drinking water source. The susceptibility modeling process can be shown graphically as in the figure below.

The models are being developed using all available existing water quality data on "raw" water for the contaminants of concern. This includes raw water monitoring data collected by the purveyor, the New Jersey Department of Environmental Protection, or an outside agency such as USGS-NAWQA. Continuous distribution plots of contaminant concentrations are made to determine which sites exceed, or are close to exceeding, drinking water standards. Spatial patterns in contaminant occurrence are assessed using maps. Specific statistical analyses are performed to detect correlations between occurrence, and hydrogeologic sensitivity, and land-use intensity. In assessing hydrogeologic sensitivity for groundwater systems, variables such as aquifer type, distance from the outcrop area, soil type, and depth to the top of the casing perforations are considered. For surface water systems, the size of the drainage area and the soil loss potential of the area are assessed. To develop the "intensity" portion of the information needed, the models are used in combination with the contaminant source inventory. The correlation between occurrence and land-use intensity (e.g., residential, agricultural, golf

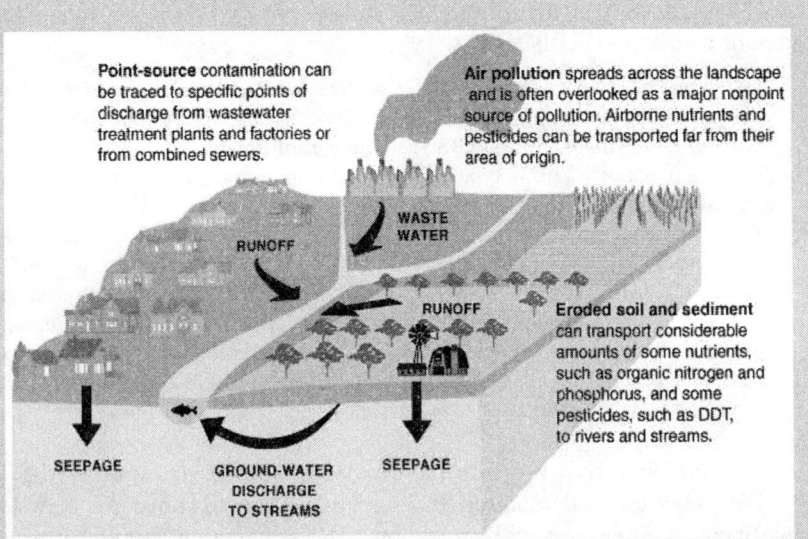

SOURCE: Eric Vowinkel, USGS.

course, industrial) is evaluated for both groundwater and surface water systems.

Based on the results of the statistical analyses combining sensitivity and intensity, a model is developed to predict the susceptibility of a water source to a particular class of contaminants. Individual sources are then given the designation "high," "medium," or "low" vulnerability. The value of conducting a susceptibility assessment for each source is to help identify which sources are more likely to be contaminated or are at risk of becoming contaminated as a result of increased contaminant release. It also identifies which types of contaminants pose a real threat to the water source, regardless of whether the contaminants are currently impacting water quality.

After the model is developed, it is then tested using existing data to determine how accurate the model predictions are. The model can then be refined as new information becomes available. The eventual goal is to be able to predict vulnerability to a particular contaminant based on land-use data.

to contamination (EPA, 1997). To assist in these efforts, EPA guidance to states typically suggests that the USGS (not necessarily NAWQA) is an important source of data and expertise (EPA, 1997):

> USGS can delineate drainage areas for surface water and contributing areas for wells. For larger drainage basins, delineations are already available in USGS hydrologic unit maps. Some GIS [geographic information system] layers are available to identify certain types of potential sources of contamination. USGS can also work with states to produce the required maps . . . USGS can use existing and new studies of watersheds, aquifers, land use, and contaminant fate and transport to determine susceptibility of drinking water sources to contamination. The USGS can also sample streams and wells to determine occurrence patterns and trends in contaminant concentrations. Such studies in Washington and New Jersey have resulted in savings, in the form of monitoring waivers, that more than covered the cost of the studies. Finally, USGS can participate in scientific review of source water protection.

Such collaborative projects have already helped to improve the cost-effectiveness of the state monitoring programs. As noted above, the New Jersey Department of Environmental Protection, after conducting a study of drinking water vulnerability with the USGS, determined that waiving monitoring for wells and intakes that are not vulnerable to pesticide contamination will save an estimated $5.1 million dollars per year (Vowinkel et al., 1996). Similarly, the Washington State Department of Health, working with local USGS personnel, obtained pesticide monitoring waivers for PWS wells that would save approximately $6.0 million dollars per year (Ryker and Williamson, 1996). Other states are joining in further collaborative efforts on formal source water assessments through NAWQA, district programs, and the USGS Drinking Water Initiative (USGS, 2001)

## Water Quality Assessments and TMDLs

One of the current strategies devised by the EPA for meeting surface water quality standards involves the application of a total maximum daily load (TMDL). A TMDL is the maximum contaminant input that a surface water body can receive and still meet water quality standards (EPA, 2000c). TMDLs are intended to be the common basis for control of point and nonpoint sources of contamination for surface waters that are designated as impaired (see more below). The TMDL provisions of the Clean Water Act are designed to provide the second line of defense for protecting the quality of surface water resources. Section 402 of the CWA established the Point Source Program, which is aimed at restricting the discharge of pollutants from municipal and industrial dischargers. The basis of the program is the National Pollutant Discharge Elimination System (NPDES) permit. Each point source must obtain a discharge permit before it can discharge wastes into surface water. The permit requires dischargers to comply with technology-based controls (uniform, EPA-established standards of treatment that apply to certain industries and municipal sewage treatment facilities) or water quality-based controls that invoke state numeric or narrative water quality standards. More than 200,000 discharge sources in the United States are subject to NPDES permits (EPA, 2000b).

Every two years under Section 305(b) of the Clean Water Act, states are required to monitor water quality and report to Congress which waters are meeting water quality standards and which are not. When technology-based controls are inadequate for water to meet state water quality standards, Section 303(d) of the Clean Water Act requires states to submit to EPA a list of impaired waters and the cause of the impairment. Once listed as impaired, a TMDL may be required as well as the development of a plan to mitigate the impairment and meet the applicable water quality standard. Developing the TMDL involves calculating the maximum amount of a pollutant that a water body can receive and still meet water quality standards, and allocating pollutant loads to the pollutant's various sources. The TMDL for the watershed is the sum of individual wasteload allocations for point sources, load allocations for nonpoint sources and natural background, and a margin of safety. At present, there are more than 20,000 such impaired water bodies identified nationally, comprising more than 300,000 miles of rivers and streams and more than 5 million acres of lakes (EPA, 2000a). Clearly, this ranges far beyond NAWQA's resources and scope, and NAWQA study units cannot be defined by TMDL needs. The top impairments from the 1998 303(d) lists are sediment, nutrients, and pathogens (EPA, 2000c). Primacy agencies, such as states, territories, and authorized tribes, are responsible for establishing and implementing TMDLs. If these agencies fail to establish the TMDLs, the EPA may be required to do it.

In brief, development of a TMDL requires prediction of the relationship between contaminant loading and the applicable water quality standard, and between source control measures and contaminant loads. This work is being undertaken largely at the state level, with the support of the EPA, but relatively few states have the resources available to provide the necessary scientific support for TMDL recommendations. Thus, an important opportunity appears to exist for NAWQA to assist the TMDL process, since TMDL assessments require the type of information that is typically produced by individual NAWQA study units. Actual monitoring data are a major component that NAWQA can contribute, both to the original assessment and to the determination of impairment, as well as to calibrate and evaluate model results. Also, the models and tools being developed by NAWQA may also prove useful to the states or primacy agencies.

If a NAWQA study unit coincides with a TMDL watershed, the NAWQA data are likely to be of value for the original assessment and development of the TMDL, particularly if the TMDL contaminant and the water quality standard reflect variables that are being actively measured as part of the NAWQA program. However even in non-study unit watersheds, NAWQA analyses of land use-water quality relationships can be useful for TMDL development through application of models such as SPARROW (Spatially Referenced Regressions on Watershed Attributes) (Preston and Brakebill, 1999). As discussed earlier, SPARROW is a regression-based model that has been used to develop location- and source-specific ("spatially referenced") estimates of nutrient delivery from a watershed and estimates of nutrient loss during in-stream transport. Although the application of SPARROW requires a substantial amount of watershed and water quality data, the modeling approach is sound and the results can be extremely useful to states in determining TMDLs for nutrients such as nitrogen and phosphorus. SPARROW can also be used to identify pollutant contributions from various sources both inside and outside the watershed (e.g., atmospheric contributions). Although additional assessments and tools may be necessary to develop the actual load allocations on sources required by the TMDL process, USGS scientists should support opportunities to use NAWQA data and analyses and the SPARROW model to assist in the development of TMDLs (see Box 7-3). The committee reiterates that TMDLs are the states' responsibility, and NAWQA resources and scientists should not be diverted to working on TMDLs beyond the data and technical assistance that they can readily provide to the states. In many cases, USGS assistance in the process should be provided through cooperative arrangements between the state and the USGS district office. Partly because TMDLs can become resource intensive and partly because the TMDL process must involve state and local policies, the committee feels strongly that TMDL development is not the general purview of NAWQA and the USGS.

## CONCLUSIONS AND RECOMMENDATIONS

The national scope of NAWQA, and its potential to provide a nationwide perspective on the status, trends, and major factors that affect water quality, make cooperation and coordination within USGS and externally, a priority to optimize the massive data collection, as well as data use and interpretation, efforts. In the committee's view, NAWQA program staff have done an excellent job of establishing cooperative relationships within USGS and with external local, state, and federal programs. For example, the USGS and NAWQA are to be commended for developing liaison staff within EPA offices. USGS staff are providing EPA important scientific perspective, for example, in technical approaches to water quality assessments, development of regional nutrient criteria, and the identification of emerging contaminants. This has provided important technical expertise to EPA but has also provided USGS with important perspectives on EPA's information needs. These efforts should be continued in Cycle II.

Other agencies that wish to utilize NAWQA or coordinate water quality-related programs with NAWQA also have a responsibility to collaborate and provide resources with staffing, fiscal, or in-kind support where possible, to help

> **BOX 7-3**
> **TMDLs—Neuse River Case Study**
>
> One current example that illustrates the opportunities and constraints associated with SPARROW applications for TMDL development is in the Albemarle-Pamlico Drainages (ALBE) NAWQA Study Unit. The Neuse River estuary does not meet water quality standards for its intended use and has been listed as impaired on North Carolina's 303(d) list. A nitrogen TMDL required for the Neuse in North Carolina is currently being developed through collaboration of the University of North Carolina Water Resources Research Institute and the North Carolina Division of Water Quality. The Neuse watershed is part of the Albemarle-Pamlico Study Unit, and the SPARROW watershed model is one of the models being used to assess the TMDL. Unfortunately, the timetable for the Neuse nitrogen TMDL is not compatible with the proposed schedule for development of a SPARROW model by USGS for the Neuse, since the ALBE study unit is not currently in its high-intensity period of monitoring. In recognition of this timing mismatch and of an opportunity for mutually beneficial collaboration, NAWQA scientists have advanced their schedule for SPARROW development on the Neuse to provide assistance to North Carolina for TMDL development. This flexibility and cooperative spirit are to be applauded. Spatial data needs (e.g., land-use and digital elevation model data) to successfully implement SPARROW are substantial; without USGS-NAWQA assistance, it is doubtful that SPARROW would be ready in time to provide assistance for the Neuse TMDL.

cover their unique needs and requests. The USGS also has to try to ensure reasonable overhead to keep transaction costs reasonable for cooperators. EPA and several states are already providing complementary or cost share funding to build on NAWQA's base and are to be commended for these efforts.

Although such cooperative efforts and design changes that can improve the efficiency of NAWQA are always warranted, providing the national perspective of the nation's waters that most policy makers desire requires adequate support. Congress should recognize this. As large as NAWQA is, program resources are often too constrained to meet its national goals. Continued flat budgets and budget cuts will not allow NAWQA to meet its goals or to provide the results that Congress and other agencies desire without continued design changes and cutbacks. Further, NAWQA cannot become dependent on other cooperative agencies for operational budget support. This will result in loss of control over its national design and loss of perspective for its national goals.

These cooperative efforts have strengthened NAWQA and have improved the visibility and viability of the USGS as a whole. However, such efforts also come at a price. Cooperation has a cost, particularly in staff time to keep effective communications in place. While cooperative efforts are valued, NAWQA cannot be all things to all people—or agencies. In the area of program coordination, the committee offers the following recommendations:

- NAWQA must stay firm in its design to meet its national goals, and should not change critical design plans to meet the diverse needs of the many federal, state, and local agencies that want to participate in the program or utilize its data and information. Thus, NAWQA must maintain its careful balancing act to uphold its design principles that draw other agencies to NAWQA, while finding ways to collaborate that build and improve NAWQA. Perhaps more importantly, such collaboration should strive to improve and strengthen other water-related programs to enhance the total knowledge of the nation's water resources.
- For the continued success of NAWQA, the USGS must continue to manage the creative tension of the joint efforts among its internal programs—NAWQA, district offices, the NRP, the Toxics Program, and NASQAN. Based on the committee's investigations and deliberations during this study, several potential areas of stress need continuing dialogue: (1) the management problems created by the fluctuation in resources and staffing that occurs between the high-intensity and low-intensity phases of NAWQA must be addressed; (2) district-level staff must fully appreciate the national goals of NAWQA, and likewise NAWQA personnel should listen to implementation concerns from their local expertise; (3) the coordinating committee for NAWQA and NASQAN should be implemented; and (4) NAWQA must continue to recognize and share full and proper credit with its partners.
- The local study unit and national liaison committees should be continued in Cycle II. NAWQA should consider providing training as well as further information sharing ("lessons learned") for study unit teams to make the local liaison efforts more beneficial for USGS and local users.
- States are for the most part, and should increasingly be, important users of NAWQA data. Several states have developed collaborative efforts (generally, jointly funded) with the USGS on many aspects of water resource assessments that have proved very beneficial to the states and water utilities. The USGS should continue to foster the use of the good science produced by NAWQA with its state cooperators. Of particular note, USGS scientists should support opportunities to use NAWQA analyses and the SPARROW model for TMDL development. TMDLs, however, are the states' responsibility. NAWQA resources and scientists should not be diverted to working on TMDLs beyond the data and technical assistance that they can readily provide to the states.

## REFERENCES

Alpers, C. N., H. E. Taylor, and J. L. Domagalski (eds.). 1999. Metal Transport in the Sacramento River, California, 1996-97. Volume 1: Methods and Data. U.S. Geological Survey Water-Resources Investigations Report 99-4286. Sacramento, Calif.: U.S. Geological Survey.

Alpers, C. N., R. C. Antweiler, H. E. Taylor, P. D. Dileanis, and J. L. Domagalski (eds.). 2000. Metal Transport in the Sacramento River, California, 1996-97. Volume 2: Interpretations of Metal Loads. U.S. Geological Survey Water-Resources Investigations Report 00-4002. Sacramento, Calif.: U.S. Geological Survey.

CADWR (California Department of Water Resources). 1993. Sacramento-San Joaquin Delta Atlas. Sacramento, Calif.: California Department of Water Resources.

Domagalski, J. L., D. L. Knifong, D. E. MacCoy, P. D. Dileanis, B. J. Dawson, and M. S. Majewski. 1998. Water Quality Assessment of the Sacramento River Basin, California—Environmental Setting and Study Design. U.S. Geological Survey Water-Resources Investigations Report 97-4254. Sacramento, Calif.: U.S. Geological Survey.

Dubrovsky, N. M., C. R. Kratzer, L. R.. Brown, J. M. Gronberg, and K. R. Burow. 1998. Water Quality in the San Joaquin-Tulare Basins, California, 1992-95. U.S. Geological Survey Circular 1159. Reston, Va.: U.S. Geological Survey.

EPA (U. S. Environmental Protection Agency). 1997. State Source Water Assessment and Protection Programs; Final Guidance. EPA 816-R-97-009S. Washington, D.C.: U.S. Environmental Protection Agency, Office of Water.

EPA. 2000a. Atlas of America's Polluted Waters. EPA 840-B-00-002. Washington, D.C.: U.S. Environmental Protection Agency, Office of Water.

EPA. 2000b. NPDES Permit Program–General Information. U.S. Environmental Protection Agency, Office of Waste Management. Available online at http://www.epa.gov/owm/gen2.htm.

EPA. 2000c. Total Maximum Daily Load (TMDL) Program: 1998 Section 303(d) List Fact Sheet: National Picture of Impaired Waters. Available online at http://www.epa.gov/owow/tmdl/ states/ national.html.

EPA. 2001. Regulatory Support Document for Hexachlorobutadiene. EPA 815-R-01-009. Washington, D.C.: U.S. Environmental Protection Agency, Office of Water.

Focazio, M. J., A. H. Welch, S. A. Watkins, D. R. Helsel, and M. A. Horn. 1999. A Retrospective Analysis on the Occurrence of Arsenic in Ground-Water Resources of the United States and Limitations in Drinking-Water Characterizations. U.S. Geological Survey Water-Resources Investigations Report 99-4279. Reston, Va.: U.S. Geological Survey.

Hellkamp, A. S., J. M. Bay, C. L. Campbell, K. N. Easterling, D. A. Fiscus, G. R. Hess, B. F. McQuaid, M. J. Munster, G. L. Olson, S. L. Peck, S. R. Shafer, K. Sidik, and M. B. Tooley. 2000. Assessment of the condition of agricultural lands in six Mid-Atlantic states. Journal of Environmental Quality 29:795-804.

Hess, G. R., C. L. Campbell, K. D. A. Fiscus, A. S. Hellkamp, B. F. McQuaid, M. J. Munster, S. L. Peck, and S. R. Shafer. 2000. A conceptual model and indicators for assessing the ecological condition of agricultural lands. Journal of Environmental Quality 29:728-737.

Hooper, R. P., D. Goolsby, S. McKenzie, and D. Rickert. 1996. National Program Framework, Part I., NASQAN Redesign, Draft. Denver, Colo.: U.S. Geological Survey.

Jones, K. B. et al. 1997. An Ecological Assessment of the United States Mid-Atlantic Region. EPA/ 600/R-97/130. Washington, D.C.: U.S. Environmental Protection Agency, Office of Research and Development.

Larson, S. J., and R. J. Gilliom. 2001. Regression Models for Estimating Herbicide Concentrations in U.S. Streams from Watershed Characteristics. In Press.

Mallard, G. E., J. T. Armbruster, R. E. Broshears, E. J. Evenson, S. N. Luoma, P. J. Phillips, and K. R. Prince. 1999. Recommendations for Cycle II of NAWQA. U.S. Geological Survey NAWQA Planning Team. U.S. Geological Survey Open-File Report 99-470. Reston, Va.: U.S. Geological Survey.

Miller, T. L., and W. G. Wilber. 1999. Emerging drinking water contaminants: Overview and role of the National Water Quality Assessment Program. Pp. 33-42 in Identifying Future Drinking Water Contaminants. Washington, D.C.: National Academy Press.

Preston, S. D., and J. W. Brakebill. 1999. Application of Spatially Referenced Regression Modeling for the Evaluation of Total Nitrogen Loading in the Chesapeake Bay Watershed. U.S. Geological Survey Water-Resources Investigations Report 99-4054. Reston, Va.: U.S. Geological Survey.

Puckett, L. J. 1994. Nonpoint and Point Sources of Nitrogen in Major Watersheds of the United States. U.S. Geological Survey Water-Resources Investigation Report 94-001. Reston, Va.: U.S. Geological Survey.

Ryker, S. J., and A. K. Williamson. 1996. Pesticides in Public Supply Wells of Washington State. U.S. Geological Survey Fact Sheet 122-96. Tacoma, Wash.: U.S. Geological Survey.

U. S. Geological Survey (USGS). 2001. The National Water Quality Assessment Program—Informing water-resource management and protection decision. Available online at http://water.usgs.gov/nawqa/docs/xrel/external.relevance.pdf.

Vowinkel, E. F., R. M. Clawges, D. E. Buxton, D. A. Stedfast, and J. B. Louis. 1996. Vulnerability of Public Drinking Water Supplies in New Jersey to Pesticides. U.S. Geological Survey Fact Sheet 165-96. West Trenton, N.J.: U.S. Geological Survey.

Welch, A. H., S. A. Watkins, D. R. Helsel, and M. J. Focazio. 2000. Arsenic in Ground-Water Resources of the United States. U.S. Geological Survey Fact Sheet FS-063-00. Denver, Colo.: U.S. Geological Survey.

Zogorski, J. S., A. Morduchowitz, A. L. Baehr, B. J. Bauman, D. L. Conrad, R. T. Drew, N. E. Korte, W. W. Lapham, J. F. Pankow, and E. R. Washington. 1997. Fuel oxygenates and water quality. Chapter 2 in Interagency Assessment of Oxygenated Fuels. Washington, D.C.: Office of Science and Technology Policy, Executive Office of the President.

# 8

# The Future of NAWQA

The final chapter of this report opens with the pretentious title, "The Future of NAWQA." While the report title and the committee's statement of task clearly indicate issues that will affect the National Water Quality Assessment (NAWQA) Program and staff in the future, they do not fully convey the potential importance of this programmatic review. In the last decade-and-a-half, NAWQA has progressed from a sound concept to a mature program of exemplary quality and importance. NAWQA has led the way to begin the critical, sound scientific assessment of the quality of the nation's waters. Because of this initial success, NAWQA now carries with it high expectations from many other federal, state, and local agencies, as well as policy makers and legislators. At this juncture, NAWQA is in a critical period, the transition from Cycle I, its first decade of nationwide monitoring, to Cycle II, its planned second decade, when many aspects of the promise of NAWQA must come to fruition.

From its earliest concept to the current plans for the future, three goals have driven NAWQA's design and development: (1) *status*—to provide a nationally consistent description of the current water quality conditions for a large part of the nation's water resources; (2) *trends*—to define long-term trends (or lack of trends) in water quality; and (3) *understanding*—to identify, describe, and explain (to the extent possible), the major factors that affect (and cause) observed water quality conditions and trends. Although their exact wording has been refined over time, these three goals are the organizing themes for NAWQA's past, present, and future. Not surprisingly, Cycle I of NAWQA focused primarily on the status goal. While this is certainly an important and often daunting task at NAWQA's scale, Cycle II must now move beyond water quality conditions of the nation and begin more substantive assessment of long-term water quality trends and further our scientific understanding of the why and how behind water quality status and trends. All three goals are a major part of the grand design of NAWQA and are central to the charge that policy makers have placed on NAWQA since its pilot-scale origins in the 1980s. Indeed, if there are substantive "opportunities to improve NAWQA," this is the ideal time to do so—when the future of NAWQA is now. Thus, the committee hopes that the timing of this report can contribute to the future of NAWQA—a future that has clearly become of widespread importance to the nation.

As discussed in Chapter 1, in concert with NAWQA leadership, the committee was asked to spend considerable time considering and discussing NAWQA Planning Team (NPT) and NAWQA Cycle II Implementation Team (NIT) planning documents and guidance reports.[1] This ultimately became the crux of the committee's focus. While the statement of task includes four particular issues (defined below), the committee decided early in the study that the operational and organizational theme for the report had to reflect the broadest charge of the statement of task—to "provide guidance to the U.S. Geological Survey [USGS] on opportunities to improve the NAWQA program." (The first component of the statement of task—to provide an initial assessment of the program's general accomplishments to date—is covered in Chapter 1.) In this regard, many of the conclusions and recommendations of this report go beyond those implied by a strict reading of the statement of task. Furthermore, in presenting such a comprehensive programmatic overview, the committee deemed it necessary to make some general comments about budgets and resources where pertinent to the scope of the proposed changes and additions to the NAWQA program.

It is important to state that the four particular statement-of-task issues are intermixed throughout the NIT plans and guidance for Cycle II, and these plans are also organized around the three major goals of NAWQA. To try to ensure its greatest utility, the committee organized much of this report around the same structure as that used in the NAWQA planning and implementation documents for Cycle II and has addressed those four particular issues in various, appropriate places throughout the report. Thus, the report's organization is essentially a blend of the NAWQA implementation plans for Cycle II and the committee's statement of task. The committee feels that this organization is both functional and logical so that the report can more directly address issues raised in NIT guidance reports.

In this chapter, the committee puts the report's findings in a broad context that includes three major components. First are the four particular issues from the statement of task and how the committee specifically addressed them throughout the report. The second component includes some of the broader concerns and conclusions throughout the report regarding the future of NAWQA. Last is the

---

[1] Most especially Gilliom, R., et al., 2000. Study-Unit Design Guidelines for Cycle II of the National Water Quality Assessment (NAWQA). U.S. Geological Survey NAWQA Cycle II Implementation Team. Draft for internal review (11/22/2000). Sacramento, Calif.: U.S. Geological Survey.

relationship of this report's findings to the conclusions of another recent National Research Council (NRC) report, *Future Roles and Opportunities for the U.S. Geological Survey*.[2] That report presents the conclusions and recommendations of the NRC Committee on Future Roles, Challenges, and Opportunities for the U.S. Geological Survey. NAWQA itself is a major program within the Water Resources Division of the USGS, and hence this committee's assessment of NAWQA is a microcosm of these larger issues. Also, some recommendations in this report likely go beyond NAWQA's responsibilities and/or capabilities, and these should be addressed to the broader programs of the USGS. In this regard, the committee places its conclusions and recommendations in the context of that broader NRC report in the hope that they will enhance the utility of both reports.

## FOUR CROSSCUTTING ISSUES

The full statement of task that was developed to initiate this study reads as follows:

> The committee will provide guidance to the U.S. Geological Survey on opportunities to improve the NAWQA program. The committee will conduct an initial assessment of general accomplishments in the NAWQA program to date by engaging in discussions with program scientists and others such as users of NAWQA products, and by reviewing USGS internal reports on opportunities to improve NAWQA. The four main activities of the study committee will then be to: 1) recommend methods for the improved understanding of the causative factors affecting water quality conditions; 2) determine whether information produced in the program can be extrapolated so as to allow inferences about water quality conditions in areas not studied intensely in NAWQA; 3) assess the completeness and appropriateness of priority issues (e.g., pesticides, nutrients, volatile organic compounds, and trace elements) selected for broad investigation under the national synthesis component of the program; and 4) describe how information generated at the study unit scale can be aggregated and presented so as to be meaningful at the regional and national levels.

As noted previously, while these four specific issues are indeed addressed by the committee, they cut across the three continuing NAWQA goals as related to Cycle II and are thus addressed throughout the report. Not only do these four issues crosscut the three goals and related components of planned Cycle II NAWQA investigations, but the issues themselves are interrelated. The following section summarizes some key findings related to these four issues. It is important to note that this chapter does not repeat or reiterate every conclusion and recommendation from this report but rather tries to provide some highlights

---

[2]National Research Council (NRC). 2001. *Future Roles and Opportunities for the U.S. Geological Survey*. Washington, D.C.: National Academy Press.

with an emphasis on more overarching issues. The reader should refer to the individual chapters and the Executive Summary for a comprehensive presentation of the report's conclusions and recommendations.

## Improved Understanding of the Causative Factors Affecting Water Quality Conditions (Issue 1)

Chapter 5 of this report is devoted to a discussion of the understanding goal of NAWQA for Cycle II and its themes and related objectives to address cause-and-effect issues. Many sound approaches proposed by NAWQA staff have already been developed in Cycle I and should provide a solid foundation for subsequent development and implementation in Cycle II—particularly in key areas such as contaminant fate and transport, groundwater-surface water interactions, and stream ecosystem studies. These are important issues that should be priorities for targeted studies, as discussed in Chapter 5. Understanding causative factors is such a broad topic that generalizations are difficult except to note the importance that modeling will have to play. As Chapter 5 states, the use of models of various kinds and complexities will be an important component of these Cycle II studies. Models, in a simple sense, can help to organize the conceptual framework (e.g., mass balance) and to organize one's thinking about an environmental system. Models represent conceptual and mathematical relationships between the observations that we interpret as cause and effect. In a quantitative sense, they can be used as tools for diagnosis and explanation of the underlying mechanisms, helping to test hypotheses associated with the fate and transport of pollutants in the environment. Standardizing some aspects of the modeling approaches used within NAWQA should also be done to facilitate comparisons among different study units (which can assist aggregation of data; see issue 4). As important as modeling can be, however, it remains a tool that does not replace field work and the importance of a rigorous field design.

As discussed in Chapter 5, many aspects of the planned Cycle II understanding studies have not yet been described in sufficient detail for the committee to definitively comment on them. Any cause-and-effect study will have unique aspects related to local conditions, hence, many details must logically come from the study units themselves to assess the appropriate topics and design for the field area. These cause-and-effect studies must, of necessity, be smaller and more focused than many other components of NAWQA. Because the resources to conduct such studies in Cycle II are likely to be constrained, they will also have to be carefully prioritized: a few, well-designed, intensive studies on a narrow topic will likely be better than diluting resources among too many study units. In this regard, the Cycle II understanding studies represent a prime area in which NAWQA may benefit from cooperative work with other federal agencies, particularly with state, local, and academic collaborators who are likely to already be involved with focused cause-and-effect studies. Such thematic studies will likely

THE FUTURE OF NAWQA

have considerable local interest (e.g., cost sharing) and perhaps considerable local expertise (e.g., joint studies), as well as possible field installations, to build on.

Despite the continuing importance that this committee (and past NRC committees) places on the appropriate use of models in NAWQA, a major recommendation is that NAWQA (and USGS water programs in general) redirect their modeling efforts. Many current modeling efforts are rapidly becoming too ambitious, complex, and overparameterized. The committee feels strongly that more focus should be placed in Cycle II on developing and using "parsimonious" models—models that are in balance with data needs and incorporate simpler mechanistic expressions and understanding of hydrogeologic systems that can be supported with available data.

## Extrapolation of NAWQA Information to Areas Outside NAWQA (Issue 2) and Aggregation of Study Unit Information to Regional and National Levels (Issue 4)

Statement-of-task issues 2 and 4 are, in part, interrelated and, as such, are discussed together. Information pertinent to these issues is discussed throughout this report. The first key assessment of these issues is presented in Chapter 2. As noted in that chapter, NAWQA is not a "statistically" designed sampling program (i.e., not based on probabilistic, random sampling theory); thus, its results are not statistically representative of the nation's waters. In other words, NAWQA results cannot simply be aggregated to provide, for example, a national mean concentration of some contaminant that can be compared over time. However, a more purely statistical design does not necessarily allow for important cause-and-effect studies or for assessment of many factors that affect water quality trends. In neither circumstance is extrapolation to unsampled areas a straightforward or easy task.

With the extensive monitoring and nationwide sample that NAWQA does collect however, it can make insightful and important statements about the status and trends of water quality throughout the nation. Yet before even this can be achieved, it is important to understand how representative NAWQA study units are of the nation's watersheds and aquifers. This was an important consideration because of the reduction from 59 (monitored and planned) study units in Cycle I to the 42 study units planned for Cycle II (see Figures 1-1 and 2-1, respectively). As discussed in Chapter 2, NAWQA has done an excellent job of iteratively utilizing hydrologic setting regions, coupled with a linear programming approach and expert judgment-based semiquantitative analyses to optimize the reduced number of study units planned for Cycle II. For this reason, the NAWQA study units planned for Cycle II are clearly representative of a "large part of the nation's water resources," but continual care must be taken to present the data correctly in these terms. Because NAWQA study units do represent such a large and significant sample of the nation, there is considerable potential for extrapolation and aggregation of their data and information. With all of the supporting and ancillary data that are collected, both regression models (e.g., SPARROW [Spatially Referenced Regressions on Watershed Attributes]) and other physical or process models that use causative factors can be employed to extrapolate to unsampled areas, with some important caveats. As discussed in Chapters 4 and 5, uncertainty in the use of such models should be assessed, as well as standard sources of variability.

As discussed in Chapter 4, NAWQA's consistent sampling regimen and analytical protocols and its integrated approach to network design provide a good baseline that minimizes many problems that can make extrapolation and regionalization problematic. As noted, the USGS has been a leader in the regionalization of hydrologic data, which has involved both aggregation and extrapolation of streamflow data in particular. NAWQA should continue to build on this experience and can adapt various statistical approaches (e.g., SPARROW) to strengthen assessments and extrapolation. To address many of these issues, as well as statistical and model issues for understanding, as recommended in Chapter 5, NAWQA should establish a statistical model support team (perhaps as a subset of the Hydrologic Systems Team [HST]) to address aggregation and extrapolation issues, including parsimonious and adaptive models, new statistical techniques (e.g., Bayesian methods, Kalman filtering), and uncertainty.

Also, issues of representativeness should continue to be explored in Cycle II. Although NAWQA has carefully documented certain factors of what is and is not represented in its watersheds and aquifers (e.g., agricultural practices, urbanization, drinking water sources), many factors are not clearly described. For example, how representative are the study units of the petrochemical and forest products industries? Such questions might have to be asked and answered for some specific data and information uses.

As discussed earlier, in relation to the understanding goal of NAWQA for Cycle II, the committee recommends some redirection of effort in model development and use. Some of the details (see Chapters 4 and 5) are also pertinent to these issues. Much of Chapter 6 is also pertinent in discussing information dissemination and policy relevance of NAWQA data and information. Through careful and consistent use of language and presentation, these data and information can be (and have already been) aggregated and made relevant to the assessment of national polices that affect water quality. It is important to state that the committee has struggled with questions about the meaning of "regional" extrapolation of data throughout its deliberations, and to its knowledge the term has never been concisely defined. By many definitions, many study units are regional (multistate) in nature; thus, their findings are pertinent to such regions. For broader regions, the same caveats and suggestions made regarding national aggregation are warranted. For example, NAWQA has well established how it represents agricultural and urban ecosystems, and comparisons at these regional levels are warranted (with the caveats that the sample is not statistically representative). As in the prior example, water quality data should not be used to directly compare conditions from areas dominated by the petrochemical industry versus forest products utilization without first establishing how representative the study units are for those characteristics.

These are also issues that can benefit from increased collaboration and cooperation. Other federal, state, and local agencies can link to NAWQA to help fill voids in unstudied areas or unstudied water quality issues. Although such agencies should already recognize this opportunity, NAWQA staff can also promote links to the program to identify and address further questions. NAWQA already has excellent examples, such as its work with U.S. Environmental Protection Agency (EPA) staff on extrapolating pesticide occurrence in surface waters or assessment of reservoirs for pesticide occurrence (see Chapters 2, 3, and 7 for further information).

## Completeness and Appropriateness of National Synthesis Priority Issues (Issue 3)

Statement-of-task issue 3 is addressed primarily in Chapter 3. The committee strongly supports the established national synthesis topics—pesticides, nutrients, volatile organic compounds, and trace elements—and commends NAWQA for its groundbreaking work in these areas. The committee also strongly supports the recent priority allocation of NAWQA staff and resources for ecological synthesis that began late in Cycle I. This represents an important area in which NAWQA can make significant contributions. In addition, the committee strongly recommends that NAWQA take a leadership role in addressing sediments in surface waters as a future national synthesis topic. This is a critical national water quality issue that NAWQA is well suited to address. Although NAWQA may not currently have the sediment data it might desire or the resources to fully address this issue, it should do the most it can with the data it has already collected and plans to collect in Cycle II, both in direct sediment measures and through its habitat and ecological assessments. Further, the USGS could also provide a leadership role for other agencies and organizations to find collaborative ways to address this important national water quality issue.

Other related priority topics and national issues are addressed in Chapters 3, 4, and 5. The committee supports the Cycle II emphasis on addressing resources important for drinking water and conducting more detailed assessment of pathogens and microbiological aspects of water quality. However, the committee strongly recommends that the USGS not get into the highly contentious area of human health risk assessment or expend resources on ecotoxicology programs. Further details on these topics are provided in the aforementioned chapters and are summarized comprehensively in the Executive Summary. In general, the committee feels that NAWQA has struck an appropriate balance in many of these water quality status areas. As discussed further below, NAWQA must struggle to proactively identify and address emerging water quality issues but must not get caught up in a reactive mode of responding to the "contaminant of the day."

## NAWQA—PAST, PRESENT, AND FUTURE

The design and management of NAWQA and its past, present, and future success represent an ongoing struggle for program balance, and this is reflected in the committee's conclusions and recommendations. As noted above, NAWQA must continue to struggle to find the appropriate balance of efforts and resources between the three primary goals of status, trends, and understanding as it enters the second decade of nationwide monitoring. The committee fully expects that NAWQA will continue to exhibit foresight and take a lead in studying emerging water quality issues and contaminants and will avoid expending unwarranted resources on a contaminant-of-the-day approach. As another exemplary balance issue, the committee notes the very important contributions NAWQA can and should make in biological assessments and ecological synthesis in surface waters, yet the committee also strongly recommends that NAWQA not embark on an ecotoxicology program (Chapter 5). This should be dealt with in collaboration with USGS's Biological Resources Division. Furthermore, although NAWQA must strive to be responsive to water quality policy and regulatory needs, it cannot be driven or controlled by these needs—thus epitomizing the struggle of doing "good science" in the public policy arena. In this regard, the committee commends NAWQA for doing an excellent job of balancing good science with policy needs in the face of flat budgets. However, NAWQA supporters, users, policy makers, and Congress itself should be made aware of the fine balancing act this requires and should be supportive of the dilemma it creates for operating such a program.

NAWQA has gone beyond the simple monitoring of regulated contaminants, such as providing important and insightful data on the occurrence of pesticide degradation products and the common gasoline additive methyl *tert*-butyl ether (MTBE), for example. While it is providing important data and tools for states to prepare their total maximum daily load (TMDL) assessments as required under the Clean Water Act (see Chapter 7), NAWQA cannot and should not prepare TMDLs for states. NAWQA is also providing important information to the non-scientific community for resources management and policy development and assessment, yet it must first and foremost stay true to its scientific design and goals. Many of these water quality findings were already known prior to NAWQA; however, NAWQA has provided a national scope and scale to issues whose importance should not be underestimated. NAWQA is providing important national water quality leadership, yet it must periodically stop and listen to the good science its collaborative partners and colleagues are developing so that the program can continue to grow and improve its service to the nation. However, as noted elsewhere in this report, NAWQA must remember that it cannot and should not seek to be "all things to all people."

NAWQA must also balance its work within the context of the agency that provides its charter, the USGS. As noted previously and discussed in greater detail below, while this committee was deliberating and writing its report, a different NRC committee published a report to help improve the USGS as a whole as it begins its third century of service to the nation. As such, this report's conclusions and recommendations should be placed within the context of that broader report in the hope that the utility of both reports will be enhanced.

## NAWQA AND *FUTURE ROLES AND OPPORTUNITIES FOR THE USGS*

The aforementioned NRC report *Future Roles and Opportunities for the U.S. Geological Survey* (see footnote 2) addresses many scientific, technical, and management issues important to the mission of the USGS as a whole. Its conclusions and recommendations seek to identify important opportunities for maintaining and strengthening the USGS. Below, the committee presents select conclusions and recommendations from that report in an italicized bulleted list and provides some summary comments pertinent to the NAWQA program. Furthermore, the committee has grouped and reordered the conclusions and recommendations for ease of discussion in this chapter.

- *The USGS is a vitally important provider and coordinator of information related to critical issues in the natural sciences.*
- *The USGS should provide national leadership and coordination in 1) monitoring, reporting, and where possible, forecasting critical phenomena; 2) assessing resources; and 3) providing geospatial information.*
- *The USGS should provide national leadership in the provision of natural resource information. This will help the United States understand its future resource needs.*

The NAWQA program is an excellent example of these conclusions and recommendations. NAWQA is providing key national leadership in monitoring, reporting, and assessing the quality of surface water and groundwater resources across the nation. Furthermore, NAWQA is playing a vital role in balancing its good science with responsiveness to policy and regulatory needs. This is a vital function. It has long been a policy maxim that good water quality monitoring is necessary to assess status, trends, and understanding and that such monitoring is best performed by a science (not a regulatory) agency. Once such monitoring is tied to regulators it becomes suspect (i.e., the analogy of the fox guarding the hen house), and regulatory monitoring typically cannot have the breadth and foresight (or often the authority) to address emerging water quality problems. Independent monitoring *and* data analyses are vital to provide unbiased input into "government performance and review." Congress and the U.S. Department of the Interior have to ensure support for such independent science, even when it reports data and information that are unpleasant or unexpected.

The other NRC committee also noted that the USGS has long provided national scientific leadership in areas such as surface water hydrology. Hopefully, in concert with NAWQA, the USGS can provide the further leadership needed in the science of water quality. Also, with the USGS as a recognized source of high-quality spatial information, it is hoped that NAWQA can strengthen the way it provides electronic data and information and further the evolution of interactive analytical information systems.

- The USGS is evolving from an agency that was organized primarily to discover what is out there, to one that tries to understand what is out there, to one that tries to understand how what is out there works (i.e., process understanding).
- The USGS should place more emphasis on multi-scale, multidisciplinary, integrative projects that address priorities of national scale.
- The USGS should emphasize system modeling as a powerful tool for integrative science.
- The USGS should develop a research agenda that is balanced appropriately between problem-specific research and core research.

NAWQA is exemplary of the evolution to assess understanding and cause-and-effect studies. NAWQA's integrated hierarchy of different levels of study, such as watersheds nested within watersheds, to aid understanding and extrapolation among scales is an example that might be studied for other purposes within the USGS. As noted previously, the committee has also endorsed the use of system modeling as an important tool for NAWQA and has recommended some redirection of model development effort. New computational resources can certainly increase model capabilities, but the committee feels that new efforts should focus on parsimonious models with more realistic data requirements that can be widely used and provide more useful information. Also as discussed previously, NAWQA is a continual struggle for balance among its three goals that can be viewed as the balance of problem-specific and core research (i.e., understanding versus status and trends).

- The USGS should develop a more effective process to assess and prioritize customer needs.
- As a major federal science agency, the USGS cannot afford to be without external advisory committees. The USGS should establish and make extensive use of external advisory committees.
- To achieve its mission goals, the USGS will have to strengthen coordination and collaboration with other federal agencies as well as with states, academia, and industry.

As discussed extensively in Chapters 6 and 7 in particular, NAWQA has made good use of external advisory groups and has done an excellent job of establishing collaborative relationships. In this regard, NAWQA may again provide some examples for the USGS to consider. However, this process is a never-ending task. Both NAWQA's use of and its approach to local (study unit) and national advisory committee's have to be strengthened and made more consistent. There has been a wide range in the quality and extent of interaction of local advisory groups, in particular. The committee also recommends that NAWQA strengthen its coordination and collaboration efforts in Cycle II, particularly to address unmet resource needs, and detailed cause-and-effect studies (including groundwater-surface water interface studies and ecosystem studies) and to develop the expertise that it does not currently have for various detailed studies. NAWQA's federal, state, local, and academic partners have much more to offer. The committee specifically recommends that the USGS and NAWQA provide leadership and bring other collaborators to address sediment in surface waters as a national synthesis topic and priority.

- Long-term databases are one of the USGS's most important contributions to the nation, and care must be taken not to disrupt them.
- As the agencies responsibilities continue to increase, its budget should be increased to a level commensurate with the tasks.
- Future demands placed on the USGS can be expected to exceed the capacity of its financial and human resources.

The water quality trend and cause-and-effect analyses that are the primary emphasis of Cycle II of NAWQA are inherently long-term databases, requiring long-term support. There is simply no other way to answer such questions as, Is the quality of water across the nation getting better or worse? This committee, and nearly all users of NAWQA with which it has interacted, recommend that NAWQA do more, not less—yet NAWQA has already exceeded its resources; NAWQA's resources have not grown to keep pace with annual inflation, and it has had to significantly redesign for Cycle II. While NAWQA has done an exemplary job of downsizing to 42 planned study units for Cycle II, it cannot continue to downsize and still be considered a *national* water quality assessment. Although it could certainly be redesigned, this would likely undo the basis for assessment of trends and waste a decade or more of effort. To address long-term trends in water quality across the nation, one must recognize the importance of these issues and the concurrent need for long-term support to allow for consistency in the data gathering and analyses efforts. The committee has noted (Chapter 5) particular concern about whether sufficient staff, expertise, and resources are available to address understanding, cause-and-effect studies. As discussed throughout this chapter, and indeed the entire report, the future success of NAWQA is itself a struggle for balance of resources and scientific endeavors in a water policy environment. Current and future demands already exceed the capacity of NAWQA, but it is the hope of the committee that policy makers, politicians, and program managers can strike the necessary balance that will allow NAWQA to continue to provide important water quality data and information for the nation.

# APPENDIXES

# Appendix A

# Extracts from *Study-Unit Design Guidelines for Cycle II of the National Water Quality Assessment (NAWQA)*[1,2]

**NAWQA CYCLE II IMPLEMENTATION TEAM**
**DRAFT FOR INTERNAL REVIEW—11/22/2000**[3]

Ken Bencala
Carol A. Couch
Lehn Franke
Dennis Helsel
Wayne W. Lapham
Jeffrey Stoner
William G. Wilber
John Zogorski

Wade Bryant
Neil M. Dubrovsky
Bob Gilliom (Chair)
Ivan James
Dave Mueller
Marc A. Sylvester
David M. Wolock

---

[1] Gilliom et al., 2000b.
[2] Extract prepared in cooperation with NAWQA/Cycle II Implementation Team (NIT) staff for exclusive use in this report. Since the structure of much of this National Research Council report (especially Chapters 3 to 5) evolved to be consistent with several iterations of the NIT report that were provided throughout the study, the committee deemed it appropriate to provide an extract of it. However, in preparing this extract, detailed budgetary information and several sections, figures, and tables of the NIT report were purposefully omitted in order to conserve space (e.g., the November 22, 2000, version was 179 pages, excluding appendixes) and editorial changes were minimal.
[3] It is also important to note that subsquent versions of this report were prepared by the NIT after the November 22, 2000, version (i.e., 06/22/01 and 06/28/01).

# NIT REPORT TABLE OF CONTENTS

INTRODUCTION
    *Themes for Cycle II Investigations*
    *Scope of Cycle II Design Guidelines*
    *Overview of Design and Organization of Report*

I. ENVIRONMENTAL FRAMEWORK FOR DESIGN
  Natural Factors
    *Hydroclimate*
    *Soils*
    *Aquifers*
  Anthropogenic Factors
    *General Land Use*
    *Urban*
    *Water Use*

II. NATIONAL TREND NETWORKS
  Overview
    *Themes and Objectives*
    *Approach and Emphasis*
    *Implementation Strategy*
  Trend Network for Streams
    *Types of Stream Sites*
    *Overview of Sampling Strategy*
    *Site Allocation and Selection*
  Trend Network for Contaminants in Sediment
    *Overview of Sampling Strategy*
    *Types of Reservoir and Lake Sites*
    *Cycle I Studies of Lake-Sediment Cores*
    *Site Allocation and Selection*
  Trend Network for Ground Water
    *Types of Ground-Water Studies*
    *Overview of Sampling Strategy*
    *Study Allocation and Selection*
    *Sub-Unit Surveys*
    *Agricultural Land-Use Studies*
    *Urban Land-Use Studies*
    *Factors Still to Be Considered That May Change Final Selections for All Land-Use Studies in the Trends Network*
    *Analysis of Reference Conditions near Land-Use Studies*

*APPENDIX A*

III. NEW STATUS ASSESSMENTS
   Overview
      *Themes and Objectives*
      *Approach and Allocation of Effort*
      *Implementation Strategy*
   New Status Assessments for Streams
      *Agricultural Sites*
      *Urban Sites*
      *Reference Sites*
   New Status Assessments for Ground Water
      *Types of Ground-Water Studies*
      *Overview of Sampling Strategy*
      *Study-Component Allocation and Selection*
   New Status Assessments for Contaminants Not Previously Sampled by NAWQA
      *Mercury*
      *New Pesticides and Degradation Products*
      *Pathogens*
   Drinking-Water Source Assessments
      *Streams*
      *Ground Water*

IV. SPATIAL STUDIES OF EFFECTS OF LAND-USE CHANGE ON WATER QUALITY
   Overview
      *Themes and Objectives*
      *Approach and Allocation of Effort*
   Streams
      *Objectives for Streams*
      *Study Design for Streams*
   Ground Water
      *Study Design for Ground Water*

V. TARGETED STUDIES OF FACTORS THAT GOVERN WATER QUALITY
   Overview
      *Themes and Objectives*
      *Implementation Strategy*
   Targeted Water-Quality Studies
      *Selection and Prioritization of Topics for Targeted Studies*
      *Topical Study Teams*
   Hydrologic Systems Analysis and Modeling
      *Hydrologic Systems Team*
      *Modeling and Analysis Tools*
      *Modeling and Analysis Resources*

Example Designs
   *Sources and Transport of Agricultural Chemicals*
   *Nutrient Enrichment of Streams*
REFERENCES

APPENDIXES
  A. *Evaluation of Cycle II Themes and Objectives*
  B. *Glossary*
  C. *Classification of Agricultural Land by Crop Groups*
  D. *Evaluation of Hydrologic Landscapes*
  E. *Analysis of Stream Networks*
  F. *Analysis of Ground Water Networks*

## INTRODUCTION

Cycle II of the U.S. Geological Survey's National Water Quality Assessment Program (NAWQA) will be the second decade of NAWQA in which 42 study units are revisited in three groups of 14 on a rotational schedule (Gilliom et al., 2000a). Similar to Cycle I, each group will be intensively studied for three years, followed by six years of low-intensity assessment. The primary emphasis of Cycle II (2001-2011) is to assess long-term trends in water quality and to improve our understanding of the factors and processes that govern water quality. The third priority is to fill critical remaining gaps in the status assessment, which was the main focus of Cycle I (1991-2001). This balance of priorities follows the recommendation of the NAWQA Planning Team (Mallard et al., 1999), which concluded:

> The primary goals of NAWQA during its first decade continue to be appropriate as the program begins Cycle II. These goals are:
>
> • Provide a nationally consistent description of current water-quality conditions for a large part of the nation's water resources. [*status*]
> • Define long-term trends (or lack of trends) in water quality. [*trends*]
> • Identify, describe, and explain, as possible, the major factors that affect observed water-quality conditions and trends. [*understanding*]
>
> To be successful NAWQA must continue to focus on all of these goals. However, there should be a shift in the relative emphasis and resources given to the three goals as the program moves into its second decade. Relative to the first Cycle, the first goal, occurrence and distribution, should receive less emphasis in Cycle II. The third goal, explanation, should receive greater emphasis. The relative emphasis given to trends should increase in Cycle II because low-intensity phase (LIP) sampling, a key component for trends analysis, was not fully implemented during Cycle I.

APPENDIX A

The purpose of this report is to describe the design and implementation strategy for Cycle II investigations in NAWQA study units.

## Themes for Cycle II Investigations

The design of Cycle II of NAWQA is guided by 12 major water-quality themes (Table 1), which are organized according to the three major goals of NAWQA water-quality assessment: (1) status, (2) trends, and (3) understanding

TABLE 1 Themes for Cycle II NAWQA Studies and Preliminary Allocation of Study Effort.[a]

| Theme | Question |
|---|---|
| **Status Themes** | |
| Resources not previously sampled | What is the quality of the most important streams and ground-water resources not sampled during Cycle I? |
| Drinking water resources | What are the concentrations and frequencies of occurrence of NAWQA target constituents in streams and ground water resources used as sources of drinking water? |
| Contaminants not previously sampled | What is the occurrence and distribution of contaminants not yet measured by NAWQA, such as pathogens, new pesticides, pharmaceutical products, high production volume industrial chemicals and others? |
| **Trend Themes** | |
| Trends and changes in status of resource | What are the trends and changes in the status of water quality since the NAWQA Cycle I status assessment and before? |
| Response to urbanization | How has water quality changed in response to urbanization? |
| Response to agricultural management practices | How has water quality changed in response to long-term changes in agricultural management practices such as tillage methods, chemical use, and crop patterns? |
| **Understanding Themes** | |
| Sources of contaminants | Identify and quantify the natural and anthropogenic sources of contaminants to surface and ground waters. |
| Transport Processes: Land surface to and within ground water | What is the relative importance of biogeochemical and physical processes in influencing the transport and transformation of surface- and *in-situ*-derived contaminants in the unsaturated zone and ground water as they are transported from land surface to shallow ground water and to underlying aquifers? |
| Transport Processes; Land surface to and within streams | How are contaminants transported—and with what losses and transformations—from land surfaces to streams and downstream to rivers, reservoirs, and coastal water? |
| Transport Processes: Ground-water/surface-water interactions | What is the role of exchanges and interactions between ground water and surface water in determining the degree and timing of contaminant levels? |
| Effects on aquatic biota and stream ecosystems | What are the effects on stream biota and ecosystems of contaminants, contaminant mixtures, habitat modifications, and other stressors, and what are the relative roles of the different stressors? |
| Extrapolation and forecasting | How can we best extrapolate (spatial dimension) or forecast (temporal dimension) water-quality conditions for unmeasured geographic areas and future conditions (after management changes), based on knowledge of land use and contaminant sources, natural characteristics of the land and hydrologic system, and our understanding of governing processes? |

[a]Preliminary allocation of Cycle II resources information removed from table.

governing factors. In describing themes and related objectives, the term "water quality" is used according to its broad definition, which includes consideration of the chemical and physical nature of water and associated particles, habitat conditions, and the composition and health of aquatic biota. Thus, for example, the factors that govern water quality include effects of contaminants on biota as well as the sources and transport of contaminants.

Cycle II themes related to the goal of understanding governing factors, which encompass the most complex and difficult objectives to address, are further organized according to a conceptual source-transport-receptor model.

**Sources** → **Transport** → **Receptors**
(*of contaminants and* (*pathways and processes*) (*exposure and effects*)
*other stressors*)

Questions organized according to this concept may involve factors in all, some, or parts of the three model components. For example, exposure may be well characterized, but effects are unknown. Or, sources may be well characterized, but factors governing transport are not. The focus of NAWQA regarding receptors is primarily on drinking water for both ground water and streams, and on aquatic biota and ecosystems for streams. The emphasis for drinking water is to characterize potential exposure, with no direct investigation of effects on humans, whereas the emphasis for aquatic biota and ecosystems is on understanding both exposure and effects.

The 12 Cycle II themes summarized in Table 1 are based on national and study-unit priorities derived from results of Cycle I studies and priorities of cooperating agencies. The themes are variably overlapping and integrated. For each theme, specific objectives have been developed that will guide Cycle II design decisions. Within each group of Cycle II study units, however, the distribution of effort among some of the Cycle II themes and objectives, particularly for the themes related to understanding the factors that govern water quality, will vary substantially and objectives will be further refined depending on the characteristics of the study units and national priorities.

Certainly other organizational frameworks for the Cycle II themes are relevant, such as by land use, drinking water and ecological effects, surface water and ground water, and so forth. There is, in fact, no single framework that can incorporate all of the different water-quality issues and scientific problems that arise. The choice of organizing themes by major program goals emphasizes the continuity of the program design from Cycle I to Cycle II and the logical progression toward better addressing the long-term goals.

Interwoven with the 12 Cycle II themes are additional, sometimes overarching, water-quality issues that cut across multiple goals and themes even though they are explicitly emphasized in one. Examples of such issues are drinking-water quality and the condition of stream ecosystems. These cross-

APPENDIX A

cutting issues are partially incorporated into one or more of the 12 primary themes, but some issues will also require separate design consideration and special attention to integration across all three program goals. Depending on the issue and its priority and characteristics, each will be addressed in somewhat different ways by Cycle II investigations.

## Scope of Cycle II Design Guidelines

The Cycle II design guidelines described in this report are intended as a national framework for designing study unit investigations and related national synthesis activities. Study designs for status and trend assessment are mainly based on the application of relatively standardized sampling components that were developed and tested during Cycle I. In general, Cycle II status objectives will be addressed by applying the same approaches used in Cycle I to new sites and study areas, or with new types of chemical and biological analyses. Cycle II trend assessment objectives will largely be addressed by resampling selected Cycle I sites and study areas to evaluate change. In general, the guidelines are quite specific in terms of site selection and sampling design for status and trend assessment, even though careful review and analysis by study unit teams is essential and will change some selections of study sites and sampling strategies.

Guidance is less specific for studies aimed at understanding factors that govern water quality. Generally, stream sites and ground water study areas selected for trend assessment also form a foundation for studies that are designed to assess governing factors and ecological effects. The detailed studies aimed at governing factors, however, also require extensive additional data collection and analysis, including a wide range of modeling approaches. For these studies, the design guidelines focus on the implementation strategy for choosing priority topics and forming topical study teams of national, study-unit, and research scientists. These topical study teams will then develop the details of study design.

The design guidelines focus mostly on the 3-year High Intensity Period (HIP) for study units, but also include consideration of the 6-year Low Intensity Period (LIP) for the trend assessment design. Each of the three groups of 14 study units that will be studied in Cycle II will have one HIP, with the balance of the decade in the LIP.

## Overview of Design and Organization of Report

The report is organized by five main chapters that explain and document the Cycle II design and implementation strategy. The organization of the chapters maintains a close correspondence to NAWQA program goals, but the order of explanation has been adapted to allow an efficient explanation of study approaches. In particular, the National Trend Networks are presented before the New Status Assessments because much of the ground work for explaining priorities for New Status Assessments is laid in the Trend Network design. Each chapter is briefly described below by chapter number and title:

**I. Environmental Framework:** This chapter is an overview of the concept and selected key elements of the environmental framework used in developing the Cycle II design. The environmental framework is the systematically characterized set of natural and anthropogenic characteristics of the national landscape that geographically define the factors that we expect to influence water quality and how it varies throughout the nation. The distribution of factors such as land use and hydrologic characteristics are used throughout the design guidelines for status, trends, and understanding studies to prioritize and select study locations and topics.

**II. National Trend Networks:** National sampling networks designed to assess long-term trends for streams, ground water, and sediment contaminants are comprised of sites distributed among a wide range of environmental settings of the nation, which are systematically sampled over time to evaluate trends and change. In addition to meeting trend assessment objectives, these sites form the foundation of data collection for other studies with additional objectives, and the design considerations for the national trend networks define some of the most critical needs for further status assessment.

**III. New Status Assessments:** New status assessments are designed to fill the most critical gaps remaining after Cycle I studies. The three types of new status assessments in the Cycle II design are: (1) those designed to fill geographic gaps in the status assessment and trend network design, (2) assessments of contaminants in drinking-water supply sources, and (3) selective assessments of contaminants that were not extensively assessed in Cycle I.

**IV. Spatial Studies of Effects of Land-Use Change on Water Quality:** Changes in water quality that are caused by urbanization and changes in agricultural development and management, particularly the status of aquatic biota and stream ecosystems, will be evaluated by gradient studies and other spatial synoptic studies in addition to the national trends network. Gradient studies and similar synoptic studies for streams take a "space-for-time" approach to assessing change, which relies on assessment of a large number of streams with different degrees of watershed urbanization or agriculture to estimate trends likely to occur in individual streams that undergo land-use changes. Space-for-time studies will also be used to assess effects of land-use change on ground water quality, but these studies will focus on evaluating changes in water quality along ground water flowpaths at various points in relation to the times that land-use changes have occurred. The spatial studies of effects of land-use change will be closely integrated with targeted studies of governing factors.

**V. Targeted Studies of Factors That Govern Water-Quality:** The complete set of Cycle II themes and objectives for understanding factors that control water quality define a broad range of scientific and management-related issues that can only partially be addressed by Cycle II investigations, even in all three groups of study units. The approach taken in the Cycle II design is to focus studies of governing factors on a limited set of the most important water quality topics within the scope of the priority themes, and to link these studies closely to related studies and data that will be collected for other parts of the Cycle II design. These carefully targeted studies will be designed and executed by topical teams formed from national synthesis teams, study units, and other scientists.

## I. ENVIRONMENTAL FRAMEWORK FOR DESIGN

The environmental framework is the systematically characterized set of natural and anthropogenic characteristics of the landscape that geographically define the factors that we expect to influence water quality, including the significance of water quality to natural systems and the uses of water by humans. Many factors that affect the sources, behavior, and effects of contaminants and water-quality conditions are common to most hydrologic systems, although in widely varying degrees of importance. These common natural and human-related factors, such as soil characteristics and land use, provide a unifying framework for making comparative assessments of water quality within and among hydrologic systems at a wide range of scales and characteristics in different parts of the Nation. Characterizing this environmental framework was an essential element of the Cycle I status assessment that cut across all individual water-quality issues and components of Study-Unit and National-Synthesis studies (Gilliom et al., 1995).

Specific factors included in the environmental framework are used to compare and contrast findings on water quality within and among NAWQA Study Units in relation to causative factors and, ultimately, to develop inferences about water quality in areas that have not been sampled. Cycle I results, for example, have enabled the development of predictive relations between estimated pesticide use on agricultural land and natural hydrologic characteristics so that pesticide concentrations can be estimated for unmeasured streams. For Cycle II design, experiences from Cycle I, improvements in national data for some natural and anthropogenic factors, and the necessity of systematically evaluating the influence of factors controlling water quality in order to meet Cycle II objectives, have placed even greater emphasis on the environmental framework to guide NAWQA studies.

## II. NATIONAL TREND NETWORKS

### Overview

The primary goals of the National Trends Networks are to systematically assess long-term trends and changes in the quality of the nation's streams and ground water and to relate observed trends and changes to probable causes. The three National Trends Networks are: (1) Trend Network for Streams, which focuses on the chemical and physical quality of water and stream ecosystems, (2) Trend Network for Contaminants in Sediment, which focuses on particle-associated trace elements and organic contaminants, and (3) Trend Network for Ground Water, which focuses on the chemical and physical quality of ground water. In addition to addressing trend assessment goals, the intensive, long-term data collection at the sites and study areas of the trend networks form the foundation for spatial studies of the effects of land-use change on water quality and for targeted studies of factors that govern water quality, which are a major emphasis of Cycle II. Because of this foundational role that the trend networks play in the Cycle II design, the requirements for these related studies are an important consideration in design of the trend networks.

*Themes and Objectives*

The themes and specific objectives addressed to substantial but variable degrees by the trends networks are listed below, including planning estimates of the percentage of Cycle II HIP budget to be allocated to each theme.

**Trends and changes in status of resource** —What are the trends and changes in the status of water quality since the NAWQA Cycle I status assessment and before?

**Objective (T1)** Determine long-term trends and changes in the concentrations of NAWQA target constituents in (a) the most important principal aquifers used for drinking-water supply and (b) recently recharged ground water upgradient of these principal aquifers in a nationally representative range of hydrologic and land use settings.

**Objective (T2)** Determine long-term trends and changes in concentrations and loads of NAWQA target constituents, and the condition of aquatic biota and stream ecosystems for (a) streams and watersheds representative of the primary hydrologic landscape and ecoregion settings present in the study units, (b) a diversity of important stream ecosystems present in study units, (c) streams and watersheds representative of agricultural, urban, reference, transitional or mixed land-use settings in the Nation, and (d) the most important streams used for drinking-water supply.

**Response to urbanization**—How has water quality changed in response to urbanization?

**Objective (T3)** Determine the effects of urbanization on the concentrations and distributions of NAWQA target constituents in ground water.

**Objective (T4)** Determine the effects of urbanization on the concentrations and distributions of NAWQA target constituents in streams and watersheds and on stream ecosystems.

**Response to agricultural management practices**—How has water quality changed in response to long-term changes in agricultural management practices such as tillage methods, chemical use, and crop patterns?

**Objective (T5)** Determine the effects of long-term changes in agricultural management practices on the concentrations and distributions of NAWQA target constituents in ground water.

**Objective (T6)** Determine the effects of long-term changes in agricultural management practices on the concentrations and distributions of NAWQA target constituents in streams and watersheds and on stream ecosystems.

## III. NEW STATUS ASSESSMENTS

### Overview

The primary goals of Cycle II status assessments are to complete essential unfinished status assessment from Cycle I and selectively expand into new status-assessment issues that become high national priorities. The main components of the Cycle II design for new status assessment are: (1) New Stream Monitoring Sites, (2) New Ground Water Surveys, (3) New Contaminant Surveys, which focus on analyses of selected contaminants not extensively assessed in Cycle I, and (4) Drinking-Water Source Assessments. The New Stream Monitoring Sites and Ground-Water Surveys primarily target water resources in environmental settings not adequately represented in Cycle I. In most cases, these studies are also candidates for addition to the National Trends Network and many will be an integral part of data collection for topical studies of factors that govern water quality.

## Themes and Objectives

**Resources not previously sampled**—What is the quality of the most important stream and ground-water resources not sampled during Cycle I?

> **Objective (S1)** Characterize the concentrations and distributions of NAWQA target constituents in principal aquifers and selected rivers and streams that were not characterized during Cycle I.
>
> **Objective (S2)** Characterize the concentrations of NAWQA target constituents in downgradient shallow ground water and streams for: (a) residential/commercial development in large metropolitan areas, and (b) the most extensive agricultural settings in the Nation.

**Drinking-water sources**—What are the concentrations and frequencies of occurrence of NAWQA target constituents in aquifers and streams used as sources of drinking water?

> **Objective (S3)** Characterize the concentrations and distributions of NAWQA target constituents in aquifers and streams that have the greatest withdrawals of drinking water.
>
> **Objective (S4)** Improve the reporting and explanation of the potential risk to human health due to the presence of contaminant mixtures that are frequently found in current or potential sources of drinking water.

**Contaminants not previously sampled**—What is the occurrence and distribution of contaminants not yet measured by NAWQA, such as pathogens, new pesticides, pharmaceutical products, high production volume industrial chemicals and others?

> **Objective (S5)** Characterize the frequencies of occurrence and concentrations of emerging contaminants in streams and aquifers that are sources of drinking water and in streams representative of the potential ecological effects from urban and agricultural land uses.
>
> **Objective (S6)** Characterize the concentrations and distributions of total and methyl mercury in streams that have the greatest potential for human exposure to mercury through consumption of drinking water or fish.

APPENDIX A

## IV. SPATIAL STUDIES OF EFFECTS OF LAND-USE CHANGE ON WATER QUALITY[4]

### Overview

The NAWQA trend networks for streams and ground water monitor trends by systematic sampling over time at carefully selected sites and study areas that represent key water resources and land uses. Although the trend networks are the only approach in the NAWQA design to assessing trends in major aquifers and rivers, they only partially address trend issues related to changes in urban and agricultural areas. In particular, effects of urban and agricultural land use on aquatic biota and stream ecosystems are not possible to adequately assess, and interpretive analysis of the causes of water-quality trends in urban and agricultural settings is generally not possible from results of the trend network alone. In order to address Cycle II trend themes aimed at water-quality changes due to urbanization and changes in agricultural land and management practices, studies of a range of spatial studies are necessary, which are closely integrated with the trend networks and the topical studies undertaken to examine the factors that govern water quality. These studies use interpretation of spatial patterns to evaluate temporal changes in water-quality factors of interest.

## V. TARGETED STUDIES OF FACTORS THAT GOVERN WATER QUALITY

### Overview

A principal goal of Cycle II is to improve explanation and understanding of the sources of contaminants, their transport through the hydrologic system, the effects of contaminants and physical alterations on stream biota and ecosystems, and implications for the quality of drinking water. Furthermore, this understanding will be the basis for predicting water-quality conditions for unsampled geographic areas (extrapolate) and future conditions (forecast). The potential scope of this explanation and understanding goal is very broad and NAWQA contributions will have to be carefully targeted to specific questions that make important contributions to the field, address the most important problems, reasonably fit within the objectives and of NAWQA, and build on Cycle I results. The approach to addressing this goal in Cycle II has two main components: (1) targeted studies of carefully chosen topics that directly relate to factors that govern water quality, and (2) the formation of a national study team focused on hydrologic systems

---

[4]Subsequent to the November 2000 version of this report, the ground water part of this section was moved to ground water trends section and the surface water part was planned to be moved to targeted studies section.

analysis which will be integrally involved with the topical study teams and national synthesis teams.

*Themes and Objectives*

Themes and objectives for understanding sources and transport of contaminants and their effects on stream ecosystems and drinking water sources comprise a complex and highly interrelated set of study requirements. The four major topical areas are sources, transport processes, ecological effects, and extrapolation and forecasting.

**Sources of contaminants**—Identify and quantify the natural and anthropogenic sources of contaminants to surface and ground waters.

>**Objective (U1) Large-scale sources**: Identify and quantify the most important large-scale natural and anthropogenic sources of selected contaminants to major streams and aquifers with mixed land use influences and contaminant sources.
>
>**Objective (U2) Urban and agricultural sources:** Identify and quantify the most important sources of selected contaminants to recently recharged ground water and to streams within urban and agricultural settings.
>
>**Objective (U3) Spatial and temporal aspects of sources:** Characterize and determine the relative importance of spatial, seasonal, and short- and long-term inter-annual variability of natural and anthropogenic sources of contaminants to surface and ground waters.
>
>**Objective (U4) Mobilizing and metabolizing sources:** Determine the relative influence of natural processes and human activities in creating contaminants or mobilizing naturally occurring contaminants.

**Transport processes: Contaminant movement from land surface to ground water**—What is the relative importance of biogeochemical and physical processes in influencing the transport and transformation of surface- and *in-situ* derived contaminants in the unsaturated zone and ground water as they are transported from land surface to shallow ground water to underlying aquifers?

>**Objective (U5) Saturated zone transport:** Examine the extent to which the concentrations of specific contaminants in surficial aquifers are related to (a) types and distributions of land use in their recharge areas, (b) distributions of ground-water residence times, and (c) physical and biogeochemical processes in the saturated zone.

APPENDIX A

**Objective (U6) Unsaturated zone transport:** Examine the influence of natural and anthropogenic factors on the concentrations and transformations of specific contaminants in the unsaturated zone, and on their flux through the unsaturated zone to shallow ground water.

**Transport processes: Contaminant movement from land surface to surface water and downstream**—How are contaminants transported—and with what losses and transformations—from land surfaces to streams and downstream to rivers, reservoirs, and coastal water?

**Objective (U7) Land surface to stream transport:** Determine how differences in watershed characteristics (e.g. soils, terrain, climate, geology) affect the transport of contaminants from watershed land surfaces into streams.

**Objective (U8) Within-stream transport:** Determine how differences in characteristics of the stream within its catchment (e.g. terrain, climate) affect the transport of contaminants along streams.

**Transport processes: Ground-water/surface-water interactions**—What is the role of exchanges and interactions between ground water and surface water in determining the degree and timing of contaminant levels and their effects on water quality?

**Objective (U9) Large-scale ground-water/surface-water interactions:** Determine the influence of ground-water quality on stream quality and the influence of stream quality on ground-water quality at stream-reach and larger scales in a broad range of environmental settings.

**Objective (U10) Small scale ground-water/surface-water interactions:** Evaluate the effects of the riparian zone, including near-stream wetlands, various land uses/land covers, and land-management practices, on exchanges of water and associated chemicals in all directions between the land surface, shallow ground water, hyporheic zone, and streams.

**Objective (U11) Hyporheic zone ground-water/surface-water interactions:** Increase understanding of the role of the hyporheic zone on the transport and fate of contaminants in both ground water and surface water.

**Effects on stream ecosystems**—What are the effects on stream biota and ecosystems of contaminants, contaminant mixtures, habitat modification or alteration, and other stressors?

**Objective (U12) Ecological effects of contaminants:** Evaluate sediment or water toxicity at sites representing the range of environmental concentrations and mixtures of contaminants. At sites found to be toxic, determine concentrations and mixtures present in sediment and/or water and screen selected biota for physiological indicators of exposure such as biomarkers, as appropriate based on contaminants present.

**Objective (U13) Ecological effects of nutrient enrichment:** Evaluate the relation between community structure and trophic dynamics among streams receiving varying levels of enrichment from allochthonous C, N, and P.

**Objective (U14) Ecological effects of habitat modification:** Characterize and evaluate the relation between stream flow characteristics, physical habitat and community structure.

**Extrapolation and forecasting**—How can we best extrapolate (spatial dimension) or forecast (temporal dimension) water-quality conditions for unmeasured geographic areas and future conditions (after management changes), based on knowledge of land use and contaminant sources, natural characteristics of the land and hydrologic system, and our understanding of governing processes?

**Objective (U15) Evaluation of empirical extrapolation and forecasting models:** Develop, evaluate, and improve empirical models for spatial extrapolation and forecasting using statistically based methods such as regression analysis.

**Objective (U16) Evaluation of deterministic extrapolation and forecasting models:** Systematically evaluate and test selected existing simulation models for their potential value in extrapolation and forecasting of water quality in streams and ground water.

**Objective (U17) Application of extrapolation and forecasting models:** Apply the most appropriate empirical and deterministic models to specific extrapolation and forecasting objectives.

## REFERENCES

Gilliom, R. J., W. M. Alley, and M. E. Gurtz. 1995. Design of the National Water Quality Assessment Program—Occurrence and Distribution of Water-Quality Conditions. U.S. Geological Survey Circular 1112. Sacramento, Calif.: U.S. Geological Survey.

Gilliom, R. J., K. Bencala, W. Bryant, C. A. Couch, N. M. Dubrovsky, D. Helsel, I. James, W. W. Lapham, M. A. Sylvester, J. Stoner, W. G. Wilber, D. M. Wolock, and J. Zogorski. 2000a. Prioritization and Selection of Study Units for Cycle II of NAWQA. U.S. Geological Survey NAWQA Cycle II Implementation Team. Draft for internal review (1/12/2000). Sacramento, Calif.: U.S. Geological Survey.

Gilliom, R. J., K. Bencala, W. Bryant, C. A. Couch, N. M. Dubrovsky, L. Franke, D. Helsel, I. James, W. W. Lapham, D. Mueller, J. Stoner, M. A. Sylvester, W. G. Wilber, D. M. Wolock, and J. Zogorski. 2000b. Study-Unit Design Guidelines for Cycle II of the National Water Quality Assessment (NAWQA). U.S. Geological Survey NAWQA Cycle II Implementation Team. Draft for internal review (11/22/2000). Sacramento, Calif.: U.S. Geological Survey.

Mallard, G. E., J. T. Armbruster, R. E. Broshears, E. J. Evenson, S. N. Luoma, P. J. Phillips, and K. R. Prince. 1999. Recommendations for Cycle II of National Water-Quality Assessment (NAWQA) Program. U.S. Geological Survey NAWQA Planning Team. U.S. Geological Survey Open-File Report 99-470. Reston, Va.: U.S. Geological Survey.

# Appendix B

# Biographical Information

**George R. Hallberg,** *Chair,* is a principal with the Cadmus Group, Inc., in Waltham, Massachusetts, conducting environmental research, regulatory analysis, and management services. Previously he was associate director and chief of environmental research at the University of Iowa's environmental and public health laboratory and at the Iowa Department of Natural Resources. Dr. Hallberg was also an adjunct professor at both the University of Iowa and Iowa State University. He is currently serving as a member of the National Research Council (NRC) Board on Agriculture and Natural Resources and has also served on the U.S. Environmental Protection Agency (EPA) National Advisory Council for Environmental Policy and Technology and on the Office of Water's Management Advisory Group. His research interests include environmental monitoring and assessment, agricultural-environmental impacts, chemical and nutrient fate and transport, contaminant occurrence and trends in drinking water, and health effects of environmental contaminants. Dr. Hallberg received a B.A. in geology from Augustana College and a Ph.D. in geology from the University of Iowa.

**Michael E. Campana** is the director of the Water Resources Program and is a professor in the Department of Earth and Planetary Sciences at the University of New Mexico. His research interests include regional hydrogeology, tropical and arid land hydrology, surface water-groundwater-aquatic ecosystem interactions, and water resources development and management. Dr. Campana has served on several previous NRC committees, including the Committee to Review the USGS National Water Quality Assessment Pilot Program. He received a B.S. in geology from the College of William and Mary, and an M.S. and Ph.D. in hydrology from the University of Arizona.

**Daniel B. Carr** is a professor in the Department of Applied and Engineering Statistics at George Mason University. He previously worked as a senior research scientist and technical working group leader of the Exploratory Data Analysis Group in the Computational Science Department of Battelle Pacific Northwest Laboratories. His research interests include regional data display, statistical graphics for data analysis, exploratory data analysis, cognostics, regression analysis, and life science and physical science applications of statistics. He received a B.A. in mathematics and psychology from Whitman College, an M.Ed. in counseling from Idaho State University, an M.S. in statistics from Oregon State University, and a Ph.D. in statistics from the University of Wisconsin, Madison.

**Lorraine L. Janus** is deputy chief of Drinking Water Quality Control and director of Watershed Field Operations for the New York City Department of Environmental Protection (NYC DEP). Previously, Dr. Janus was employed as assistant to the senior scientist at the Canada Center for Inland Waters and a senior environmentalist for the South Florida Water Management District. Her professional activities include primary and co-authorship of the Canadian Contribution to the Organisation for Economic Co-operation and Development (OECD) Program on Eutrophication, several NYC DEP reports on phosphorus loading and impacts on water quality, water quality monitoring, and watershed monitoring and protection. She received a B.Sc. in biology and an M.Sc. in limnology from the University of Waterloo and a Ph.D. in limnology from McMaster University.

**Judith L. Meyer** is a research professor in the Institute of Ecology and director for science of the River Basin Science and Policy Center at the University of Georgia. Her expertise is in aquatic ecology, especially nutrient cycling in streams, the role of riparian zones in controlling nonpoint source pollution, and the effects of land-cover changes on stream biodiversity. She has served on several NRC committees and is a former member of the Water Science and Technology Board. She received a B.S. from the University of Michigan, an M.S. from the University of Hawaii, and a Ph.D. in ecology and evolutionary biology from Cornell University.

**Kenneth H. Reckhow** is a professor of water resources at Duke University and is the director of the University of North Carolina Water Resources Research Institute. Dr. Reckhow's research interests focus on the development, evaluation, and application of models for the management of water quality. In particular, he is interested in the effect of uncertainty on model specification, parameter estimation, and model applications. Recent work has expanded this theme to consider the effect of scientific uncertainties on water quality decision making. He recently chaired the NRC Committee to Assess the Scientific Basis of the Total Maximum Daily Load Approach to Water Pollution Reduction. Dr. Reckhow received a B.S. in engineering physics from Cornell University and an M.S. and Ph.D. in environmental science and engineering from Harvard University.

**Marc O. Ribaudo** is the leader of the Water Quality Research Program and the Coastal Resources Research Program of the Resource Economics Division, Economic Research Service, U.S. Department of Agriculture (USDA). He is responsible for developing and carrying out a research program on the broad set of water quality issues affecting agriculture and on the linkages between agricultural production and the quality of coastal resources, respectively. Specific expertise includes assessing the economic impacts of agricultural production on water users, assessing the performance of USDA conservation policies, and assessing the benefits and costs of federal and state water quality laws. He received a B.S. in natural resource management and an M.S. in agricultural and resource economics from the University of Maine. Dr. Ribaudo received a Ph.D. in agricultural economics from the Pennsylvania State University.

**Kenneth K. Tanji** is a professor emeritus in the Department of Land, Air and Water Resources at the University of California, Davis. His research focuses on water quality aspects of irrigation and drainage including mass balance of salts and trace elements, use of marginal quality waters, and soil and water chemistry in croplands, agroforestry, evaporation ponds, and wetlands. Dr. Tanji has served on previous NRC committees assessing soil and water quality and the U.S. Department of the Interior's National Irrigation Water Quality Program. He received a B.A. in chemistry from the University of Hawaii; a B.S. and M.S. in soil science from the University of California, Davis; and a D.Sc. in agricultural science from Kyoto University, Japan.

**Richard M. Vogel** is a professor in the Department of Civil and Environmental Engineering at Tufts University. His primary expertise is in the area of water resource engineering with emphasis on hydrologic, hydraulic, and statistical methods for analyzing water resource systems. His current research focuses on the areas of watershed modeling and management, regional hydrology, and environmental statistics. In addition to his academic experience, he has several years of consulting experience in the field of water resource engineering. He received a B.S. in engineering science and systems and an M.S. in environmental science and hydrology from the University of Virginia. Dr. Vogel received his Ph.D. in water resource systems from Cornell University.

**Marylynn V. Yates** is a professor of environmental microbiology in the Department of Environmental Sciences and associate executive vice chancellor at the University of California, Riverside. Dr. Yates conducts research in the area of water and wastewater microbiology. Current research focuses on contamination

APPENDIX B

of water by human pathogenic microorganisms, especially through the use of reclaimed water and biosolids; developing and improving methods to detect microorganisms in environmental samples; persistence of pathogenic microorganisms in the environment; and efficacy of water, wastewater, and biosolids treatment processes to inactivate pathogenic microorganisms. Dr. Yates previously served on the NRC Committee on Groundwater Recharge. She received a B.S. in nursing from the University of Wisconsin, Madison, an M.S. in chemistry from the New Mexico Institute of Mining and Technology, and a Ph.D. in microbiology from the University of Arizona.

## STAFF

**Mark C. Gibson** is a staff officer at the NRC's Water Science and Technology Board (WSTB) and was responsible for the completion of this study. After joining the NRC in 1998, he first supported then directed the Committee on Drinking Water Contaminants that released three reports, culminating with *Classifying Drinking Water Contaminants for Regulatory Consideration* in 2001. He is also the study director for the Committee to Assess the Services and Values of Aquatic and Related Terrestrial Ecosystems and co-study director of the Committee on Indicators for Waterborne Pathogens. Mr. Gibson received his B.S. in biology from Virginia Polytechnic Institute and State University and his M.S. in environmental science and policy in biology from George Mason University.

**Laura J. Ehlers** is a senior staff officer for the Water Science and Technology Board of the National Research Council. Since joining the NRC in 1997, she has served as study director for eight committees, including the Committee to Review the New York City Watershed Management Strategy, the Committee on Riparian Zone Functioning and Strategies for Management, and the Committee to Assess the Scientific Basis of the Total Maximum Daily Load Approach to Water Pollution Reduction. She received her B.S. from the California Institute of Technology, majoring in biology and engineering and applied science. She earned both an M.S.E. and a Ph.D. in environmental engineering at the Johns Hopkins University. Her dissertation, entitled RP4 Plasmid Transfer Among Strains of *Pseudomonas* in a Biofilm, was awarded the 1998 Parsons Engineering/Association of Environmental Engineering Professors award for best doctoral thesis.

**Ellen A. De Guzman** is a research associate at the NRC's Water Science and Technology Board. She received her B.A. degree from the University of the Philippines. She has supported a number of studies including most recently the Committee to Review the New York City Watershed Management Strategy, Committee on Drinking Water Contaminants (Phase II), Committee on Risk-Based

Analyses for Flood Damage Reduction, and the Committee to Assess the U.S. Army Corps of Engineers Water Resources Project Planning Procedures. She co-edits the WSTB newsletter and manages the WSTB homepage.